"David Gibson's groundbreaking work is a real eye-opener for all of us, music professionals and casual listeners alike. He has single-handedly rethought the whole metaphor for the visual representation of recorded music and conjured up a brand new way to interact with it. It's high time we took a new look at the antique user interfaces employed by typical MIDI and same editing tools – and this book is a great place to start." – Thomas Dolby Robertson

David Gibson uses 3D visual representations of sounds in a mix as a tool to explain the dynamics that can be created in a mix. This book provides an in-depth exploration into the aesthetics of what makes a great mix. Gibson's unique approach explains how to map sounds to visuals in order to create a visual framework that can be used to analyze what is going on in any mix.

Once you have the framework down, Gibson then uses it to explain the traditions that have been developed over time by great recording engineers for different styles of music and songs. You will come to understand everything that can be done in a mix to create dynamics that affect people in really deep ways.

Once you understand what engineers are doing to create the great mixes they do, you can then use this framework to develop your own values as to what you feel is a good mix. Once you have a perspective on what all can be done, you have the power to be truly creative on your own – to create whole new mixing possibilities.

It is all about creating art out of technology. This book goes beyond explaining what the equipment does – it explains what to do with the equipment to make the best possible mixes.

David Gibson has been teaching, engineering, and producing groups in major 24 track studios since 1982 and is the founder of Globe Institute of Recording and Production, which offers classes in San Francisco and online. He has done recording for James Brown's band, Bobby Whitlock (Derek and the Dominoes), The Atlanta Rhythm Sections, Hank Williams Jr.'s band, members of the Doobie Brothers, Lacy J. Dalton, and Herbie Hancock. He is also the co-author of the top-selling book on producing, *The Art of Producing*. Gibson is currently the #1 seller of Sound Healing music, also used in hospitals across the US, and he is the author of the best-selling book, *The Complete Guide to Sound Healing*. Gibson is the inventor of the patented Virtual Mixer mixing plug-in and a Virtual Reality healing system where you can place sounds inside a 3D image of the body.

The Art of Mixing

A Visual Guide to Recording,
Engineering, and Production

Third Edition

David Gibson

Routledge
Taylor & Francis Group

NEW YORK AND LONDON

Third edition published 2019
by Routledge
52 Vanderbilt Avenue, New York, NY 10017

and by Routledge
2 Park Square, Milton Park, Abingdon, Oxon, OX14 4RN

Routledge is an imprint of the Taylor & Francis Group, an informa business

First edition published by Mix Books, an imprint of artistpro.com LLC 1997
Second edition published by Thompson Course Technology PTR 2005

Library of Congress Cataloging-in-Publication Data
Names: Gibson, David, 1957– author.
Title: Art of mixing : a visual guide to recording, engineering, and
production / David Gibson.
Description: Third edition. | New York, NY : Routledge, 2019. |
Includes index.
Identifiers: LCCN 2018028427 (print) | LCCN 2018029414 (ebook) |
ISBN 9781351252218 (pdf) | ISBN 9781351252201 (epub) |
ISBN 9781351252195 (mobi) | ISBN 9780815369479 (hardback :
alk. paper) | ISBN 9780815369493 (pbk. : alk. paper) |
ISBN 9781351252225 (ebook)
Subjects: LCSH: Sound recordings—Production and direction.
| Popular music—Production and direction.
Classification: LCC ML3790 (ebook) | LCC ML3790 .G538 2019 (print) |
DDC 621.389/32—dc23
LC record available at https://lccn.loc.gov/2018028427

ISBN: 978-0-8153-6947-9 (hbk)
ISBN: 978-0-8153-6949-3 (pbk)
ISBN: 978-1-351-25222-5 (ebk)

Typeset in Frutiger
by Florence Production Ltd, Stoodleigh, Devon, UK

This book is dedicated to all those who just want
to know how to make it sound better.

Contents

Visuals

This book has been designed to answer the elusive questions, "What makes a great mix?" and "How do you go about creating a great mix?" Although most people know what they like, they often don't know how to achieve what they want when they're in the studio.

To answer these questions, I introduce and use visual representations of sounds as a tool for understanding the whole world of dynamics that an engineer can create with the equipment in the studio. This visual framework has now become an established tool in the Industry. Colleges around the world use the visuals to explain mixing theory.

It's easy to learn the function of each piece of equipment in the studio; you can read user's manuals or the tons of good books available that explain the equipment. The difficulty lies in knowing how to use the equipment and learning what combinations of equipment are used to create great sounding mixes. Once you know what the knobs do, which way do you turn them?

In other fields of art, there is no shortage of books that attempt to explain the whole world of aesthetics. From music to painting, scholars have tried answering the question, "What makes great art?" But recording is a relatively new field, and very little has been written about the aesthetics of mixing.

This is one of the first books to explain the aesthetic side of creating a great mix. This is no simple feat, as there are many musical styles based on any number of different instruments, all of which are recorded differently. Each style of music has its own world of values that are changing constantly. The number of variations is endless. Perhaps no one has attacked this complex subject of mixing due to the lack of a framework to analyze the process. Without a framework, it is difficult to explain what is going on and hard to remember all the different things that can be done in a mix. In the field of music, music theory provides this framework. This book introduces a framework for understanding all the dynamics that can be created in a mix.

The primary goal of this book is to give you a perspective on how the equipment works together to create every mix in the world. Once you have a perspective on what can be done, you can be truly creative on your own.

It has been said that there are no rules when it comes to recording. However, in the recording industry, there are absolutely certain high-level values that are commonly held. We know this

Visual 1.
Sound Imaging of
Instruments

because there are certain professional engineers who can create a great mix every time they sit in front of a console. These engineers command exorbitant fees because they are capable of coming up with something that most people perceive as great, every time. So what is it they are doing? They are not performing magic. They are only doing some very specific things with the available mixing tools. If you could simply understand and learn what they do, you could start down the path to becoming a great engineer. Once you have a map and you know where you're going, you'll get there much faster! And once you understand what the successful engineers are doing, you can create your own style. This book will help you develop and recognize your own values through visuals, because visuals help us to remember. After all, a picture is worth a thousand sounds.

This book will help you discover the high-level values that major engineers have and help you do the most difficult job of all: make art out of technology. This book provides the missing link between technology and aesthetics. Using the visual framework, for the first time, we can see all that goes into making a good mix, and we can begin the lifelong exploration in detail. This is the art of mixing.

Visual 2.
Structuring a Mix

What I Have Learned from the Previous Editions

Since I first wrote this book, a wonderful thing has happened. Now that this framework has become commonly accepted in the field, many have written about how this has helped them, but it has also given people a way of expressing and discussing dynamics in mixing. It has opened up a whole new world of creative exploration. Therefore, I have been receiving a huge number of ideas that have continued to expand the creative landscape of possibilities.

But more importantly, it has opened up a whole new world for myself. Using the visual framework, I have been able to expand my repertoire of recording techniques immensely. Now whenever I hear something on the radio or a new CD, I am able to immediately recognize everything that the engineer did to create that mix. Because of this, I now have a better perspective than ever on how every dynamic might be used in different types of music and songs – and it gets deeper and more complex every day.

It is my hope that this perspective can be transferred to you as it has been for thousands of others. It is a very powerful tool.

> It's really not so much about
>
> me showing you a **few dozen** techniques,
>
> but about providing you with a framework
>
> that you can use to discover
>
> **hundreds** of techniques hidden in the types of
>
> music, songs, and mixes that you like – then you can
>
> use them in your own mixes appropriately.

But now a whole new realm has opened up for me. Besides Audio Recording, we now teach Sound Healing and Therapy at my College, Globe Institute. I have been studying how sound and music affect a person physically, mentally, emotionally, and spiritually for over 15 years. We have put on seven international sound healing conferences and we also have a sound healing therapy center.

What this has done is it has given me a whole new perspective on how to make music and recordings that touch people on a much deeper level. I now have over 100 CDs that are helping people with a full range of issues including relaxation, pain management, sleep, ADD/ADHD, PTSD, depression, anxiety, grief, autism, opening the heart, and accessing higher states of consciousness. This wider perspective has been helpful not only for so-called "healing" music, but also for any style of music that I may be working on.

There are many detailed aspects that can be very powerful. These include:

- Tuning to auspicious concert pitches (like 432 Hz).
- Using ancient tuning systems (like just intonation or Pythagorean).
- Relating to frequencies based on chakras.
- Using binaural beats for brainwave entrainment.
- Being aware of how different timbres and instrument sounds affect us physically and emotionally.
- Using a slow fade on the home note of the key of the song to leave people in a profound state of peace.

- Tuning the tempo to the key of the song, and even tuning the key of the song to a person's Soul note.
- Holding an intention with 100% focus throughout the recording and mixing. It has been shown scientifically that a consistent intention affects people in very powerful ways.
- Tuning the tempo to your breath – particularly while holding a specific intention.

These are just a few of our secret weapons for making music and recordings move us in even more profound ways than it already commonly does.

It brings us back to the core aspect of what we are doing . . .

> Creating musical recordings
>
> that have
>
> Meaning
>
> more than at the mental level.
>
>
> We all know how magical music can be.
>
> We are just now learning the keys as to why . . .
>
> so we can use those keys
>
> to open up deeper levels of bliss, intensity, ecstasy and peace
>
> than ever felt before.
>
>
> And yet
>
> we've only just begun
>
> understanding
>
> how it all works.

Acknowledgments

There is a wide range of people who helped me along the way to this point where I am writing this book. As with all information, over the years, I have simply gathered together a large amount of information from a huge number of contacts and sources – and then there are those divine inspirations.

First, I probably would have never gotten into this business without the suggestion of my brother Bill. He was the first to say, "Ever thought about being a recording engineer?" Then, there were my various music instructors and all of my recording instructors, including Bob Beede and John Barsotti. There was also Herbert Zettl, whose book on video aesthetics helped to inspire the structure of this book. Craig Gower was also another inspiring force in learning about working with music. And then there was Chunky Venable who was kind enough to have the faith in me to run his studio even though I was such a newbie at that time. Much thanks goes to my producer friend, Ken Kraft, with whom I learned many of the techniques within.

There are also those various artists, engineers, and producers who have influenced my values on this long road. Everybody from Pink Floyd to Bob Clearmountain have made a huge impression on my recording and mixing values.

A very special thanks goes out to my dear friends Steven Rappaport and Ginger Lyvere, who were there at the inception of this book. They spent many hours looking over rough manuscripts, brainstorming over various ideas and concepts, and encouraging me to go forth with the project.

I would especially like to thank those who spent the time to read through earlier versions of the book at no cost: David Schwartz, Charlie Albert, Fred Catero, Roy Pritz, Bob Ezrin, and Thomas Dolby. A special thanks to Fred for his exuberant and detailed critique of the book. Extra thanks to Thomas for his evangelical support of the whole visual paradigm.

Thanks are in order to Todd Stock, who has helped with editing and been a spiritual advisor of sorts. Thanks to Archer Sully, who has been helping to bring into reality an actual working proto-type of the Virtual Mixer. Warm thanks go to Donna Compton and Patrice Newman for helping with the rough editing of the book and critiquing of the visuals. Special thanks to Donna for all her kind and caring support over the years. Thanks go to Fred Mueller for much of his graphics work on the book. Extra thanks go to Melissa Lubofsky for her visual inspirations and dedicated work in creating many of the graphics, as well as her patient efforts in helping me to learn Photoshop so I could do many of the visuals on my own. Thanks also to Alon Clarke for his enthusiasm and creativity with the photography.

Much appreciation goes out to all of the gang at ArtistPro Books for their extremely helpful, creative, and professional support: Mike Lawson, George Petersen, Lisa Duran, and Linda Gough. Special thanks to Mike Lawson for believing in the "bubble book" when others thought it was "woo-woo."

And finally, I would be amiss if I didn't thank all of my students from over the years for the innumerable suggestions and inspirations that they have brought to me. They are the real reason for this book. May this book begin an ongoing dialog about what makes a great mix, to light the way for students in the future, so none of us are "all mixed up" ever again.

Introduction

Chapter 1: All Aspects of a Recorded Piece of Music

This book is designed to explain how to create great mixes. However, the mix is only one aspect of what makes a great recording. Other factors also contribute to what is perceived as a quality recording and mix. The purpose of this chapter is to put everything that goes into making a quality recording into perspective. "All Aspects of a Recorded Piece of Music" identifies and defines each of the thirteen components of a great recording: intention, concept, hooks, melody, rhythm, harmony, lyrics, density of arrangement, instrumentation, song structure, performance, quality of the equipment/recording, and the mix. We'll then explore what the recording engineer can do to help refine each of these components. My book, "The Art of Producing," is dedicated to helping refine each of these components in more detail.

Each of these thirteen aspects must be at a minimum level of good quality. The overall quality of the recording is only as good as the weakest link. The mix is only one of the thirteen aspects, but it is one of the most powerful because it can hide some of the weaker aspects, highlight the magic in the stronger aspects, or create its own magic. The rest of the book then focuses on all of the fine details of what can be done with the thirteen aspects – the mix.

Chapter 2: Visual Representations of "Imaging"

"Visual Representations of 'Imaging'" introduces the visual framework for representing "imaging," the apparent placement of sounds between the speakers.

Section A shows the difference between the perception of physical sound waves coming out of the speakers and the imagined perception of imaging. This is important because the two are sometimes confused, and the visuals represent only imaging, not physical sound waves.

Section B, "The Space Between the Speakers," introduces the visual mapping of volume, frequency, and panning, and defines the boundaries of imaging (the limited space where a mix occurs between the speakers).

Section C continues with the mapping of audio to visuals and explains the precise considerations that were used to determine the size, color, and shape of different sounds and effects. It explains how volume, frequency range, delays, and reverb affect the amount of space used up by a sound. The section explains how you can place and move sound images throughout the 3D space between the speakers using volume, panning, and equalization. The section then uses the sound images to introduce the different structures of mixes that can be created in the studio.

Chapter 3: Guides to a Great Mix (Reasons for Creating One Style of Mix or Another)

"Guides to a Great Mix" explains all of the components to take into consideration when designing a mix. If you follow what the music and song are telling you to do, the mix will be more cohesive and powerful, and it will more clearly express what the song is all about.

Section A explains how the style of music affects the way a song is mixed.

Section B describes how the song and all of its details (the thirteen aspects) dictate the way a song is mixed. It also explains how each aspect might affect the placement of volume, EQ, panning, and effects.

Section C discusses how the dynamic mix of people involved – the engineer, the band, the producer, and the mass audience – affects the way a song is mixed. It explains the engineer's role in balancing the values of everyone involved. The most difficult job of all is to take the values, suggestions, and ideas of everyone involved in a project, decide which ones are best for the project, then diplomatically work with everyone to bring about the best recording and mix possible.

I have sometimes found that if I have walked on anyone's toes (or ego), or upset anyone along the way . . . that they will not like the mix no matter how great it is. The truth is that the energy of the people during the recording gets embedded in the music itself. Therefore, I found it is critical to work together in kind and considerate ways when expressing intense caring for wanting something to be a certain way.

Chapter 4: Functions of Studio Equipment and Visuals of All Parameters

"Functions of Studio Equipment and Visuals of All Parameters" utilizes the images outlined in Chapter 2 to describe the function of each piece of studio equipment in the mix. It briefly, but technically, explains what each piece of equipment does. (As you will see, the visuals make the details of complex functions easily understandable.)

Section A explains the basic functions of faders, compressor/limiters, and noise gates and how to set them for different instruments in various styles of music and songs.

Section B explains the differences between various types of equalizers – graphics, parametrics, and rolloffs – and describes all the frequency ranges found in sounds. This section also covers the mathematical harmonic structure of the individual frequencies that make up all sounds, or *timbres*. These harmonics are the basic building blocks of sounds. Understanding the harmonic structure is critical to understanding why an equalizer works differently on different sounds. When you use an equalizer, you are changing the volume of the harmonics in that sound, and every sound has different harmonic content. Most importantly, this section gives you a step-by-step process for using an equalizer to make something sound good – or just the way you want it to sound!

Section C covers the basics of left to right placement or panning in a mix.

Section D describes each of the common functions and parameters of delays, flangers, choruses, phase shifters, reverbs, harmony processors, and pitch correctors.

At this point, all of the basic functions of the equipment, and how each parameter is mapped out visually will have been covered. Now, when we use the visuals to show how all the equipment can be used together to create different styles of mixes, you will understand all of the fine details of the visuals.

Chapter 5: Musical Dynamics Created with Studio Equipment

"Musical Dynamics Created with Studio Equipment" explains the missing link between the dynamics created by the equipment and the dynamics in music and songs. I'll first discuss the incredibly wide range of possible dynamics that different people perceive in music, including feelings and emotions; thought forms; psychological, physiological, and physical reactions; visual imagery; and cultural and spiritual connotations.

I'll then embark on an in-depth survey of how each piece of equipment in the studio can be used to create musical and emotional dynamics. There are four main types of mixing tools that can be used to create dynamics: volume, EQ, panning, and effects. I then outline and define three levels of dynamics that can be created with the four tools: individual placement, overall patterns, and movement (changing settings).

I go into detail on how common instrument sounds have traditionally been used to create dynamics at each of the three levels.

Learning the traditions that have come to dictate the placement and movements of sounds in a mix for different styles of music helps you to make the mix better fit the song and style of music.

This is especially important because some of these traditions are very strict for certain styles of music like country or classical music. And, often the people doing these styles of music subscribe to these traditions very strictly.

On the other hand, once you know these traditions you can then push the limits of what is acceptable (and change the world bit by bit). If you go too far off the deep end of creativity, people may not accept it.

The coolest part is when you learn the traditions, and are working with someone who is a free creative spirit and you are able to mix traditions of mixing for different genres of music in order to create a whole new creative mix that no one has ever imagined. Once you can see all the possibilities, you can then put them together in new and unique ways.

Chapter 6: Styles of Mixes

"Styles of Mixes" is an exploration of the incredibly powerful dynamics that can be created when you use all of the equipment together to construct an overall style of mix. It explains how you can build "high-level dynamics" by combining multiple settings using different mixing tools. When all of the equipment is used to create similar emotional dynamics, you can produce some very powerful overall mixes.

Once you have formed a context or a particular style of mix, the most intense dynamic is to completely change all of the settings on all of the equipment at once to create a completely different type of mix or context. There is nothing so intense and powerful when it comes to engineering. This chapter discusses that technique.

Chapter 7: Magic in Music, Songs, and Mixes

"Magic in Music, Songs, and Equipment" explores the variety of possible relationships between the dynamics you can create in a mix and the dynamics that are found in music. Does the equipment enhance or cause tension with what is going on in the music and song? Most importantly, does it fit? This chapter is designed to set you on your way in this lifelong exploration of all of the relationships between mix dynamics and song dynamics. After all, relationships are what it's all about.

At this point, you will have a framework for understanding and remembering all that can be done in a mix. Then, by asking yourself if you like what they are doing, you will develop your *own* style and you can confidently do whatever *you* want.

Chapter 8: 3D Sound Processors and Surround Sound Mixing

Surround sound mixing has been used in movies for years, and is now becoming widely accepted as a format for listening to music. "3D Sound Processors and Surround Sound Mixing" takes all of the concepts discussed in this book and applies them to the use of 3D sound processors and mixing in surround sound. Visuals are especially useful in exploring all of the possibilities in this evolving mixing format.

Chapter 9: Mixing Procedures

"Mixing Procedures" details the step-by-step processes that are required to complete a mix.

The first section, "The Mixing Process" takes you through a procedure for building a mix. The second section, "Automation," explains the functions and use of automation, along with details on how to go about it. The third section, "Mastering," covers all that goes into applying the final touches to your stereo master recording, before pressing your hit CD.

Appendices

The appendices include outlines of the EQ and mastering processes for quick reference. They also include all of the homework exercises from the book as well as an explanation of the Virtual Mixer that uses the visuals in the book as a 3D interface for mixing.

The color visuals are representations of particular moments in the mix. In order to represent a true mixing process, they would be flashing on and off to the music. Therefore, some of the visuals may look busier than the mix really is.

Of course, every song has its own personality and is mixed based on that. Therefore, don't assume that there is only one way to mix any style of music. These visuals are only a reference point from which you can begin to study what is done in mixes for various types of music.

With all of this in mind . . . enjoy.

Visual A. Electronica Mix

Generally a pretty busy mix with an 808 boom loud and out front. Note the fattening on the bass, and the doubling on each of the keys. Note the delay on the synth and hi-hat. Especially unique is the doubling of the hi-synth with another instrument. The super high strings are flanged for a subtle, spacey effect. The snare is not really too loud in this particular mix.

Visual B. Blues Mix

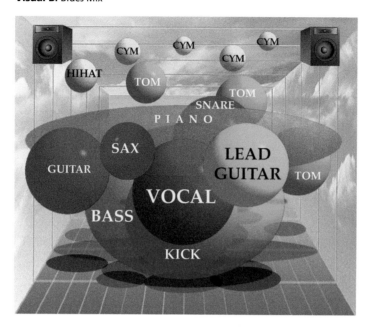

Generally a pretty clean, clear, out front mix. Note the bass is quite loud overall, with the kick drum not far behind. The rhythm guitar, the xax, and especially the lead guitar are right out front. The vocal is set back in this mix a bit, but this is not always the case. The piano is set further back a bit in the mix, but is spread in complete stereo. The toms, hi-hat, and cymbals are all set back a bit, and the snare is a bit low in this mix, which is not necessarily typical of blues.

Visual C. Rap Mix

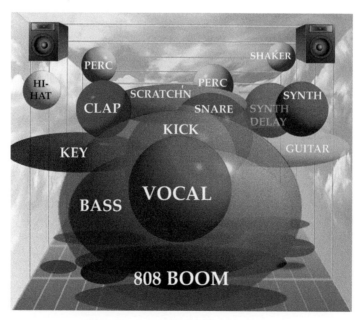

The rap mix commonly has the 808 boom boom'n and a loud vocal (though this does vary). The key, guitar, and scratch'n are all spread in stereo with fattening. Note the extremely loud clap and hi-hat; the kick is also right out front. In this mix, the snare is back a little. Also cool is the delay on the synth panned next to it. Finally, note the shaker panned opposite the hi-hat.

Visual D. Reggae Mix

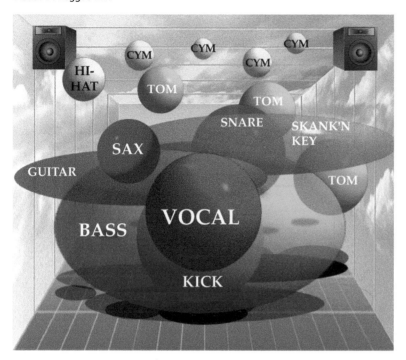

These days, reggae mixes tend to have a huge bass with the kick not far behind and the vocal right out front. Note the clarity of the sax. Both the guitar and the skank'n key are spread in stereo so they overlap a large amount. The snare is set back a little but not always, and the hi-hat is commonly right out front.

Visual E. Heavy Metal Mix

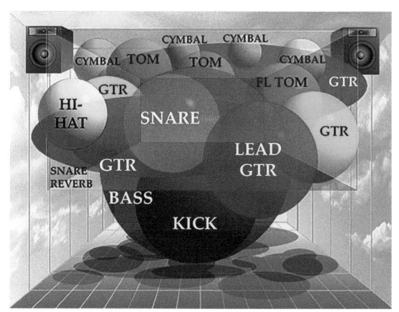

A very full arrangement and mix. Note the clarity of the low end (kick and bass), even though it is an extremely busy mix. The hi-hat, snare, and especially the lead guitar are right out front. Note the multiple guitar parts with a few panned in stereo. The reverb is present but not so loud that it muddies everything. There isn't much room left for very many effects unless there is a breakdown section in the song. The overall effect is a massive, powerful wall of sound.

Visual F. New Age Mix

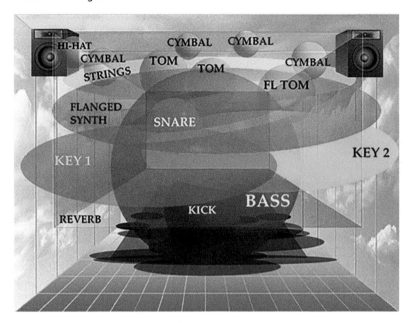

Generally, the mix is extremely full with nothing too sharp or cutting (although often individual lead sounds can be very strong). Note the fattening on the keys and strings, filling out the space. The stereo flanged synth is quite prominent here, and the bass is huge in this particular mix. The low end is kept nice and clean here, and the high-frequency cymbals placed above it all.

Visual G. Alternative Rock Mix

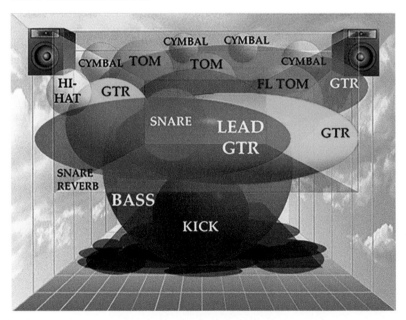

Quite full with lots of fattening and overlapping sounds. The lead guitar is spread in stereo with a rhythm guitar behind it and another stereo guitar in the background. A nice, clean low end, even though the mix is full. The kick and bass are quite present.

Visual H. Acoustic Jazz Mix

Note the incredible cleanliness and clarity of the overall mix. The bass is panned to the right and doesn't have much high end (not always the case). The guitar is right out front with the piano and the hi-hat. The kick is quite loud here, which is not typical.

Visual I. Folk Music Mix

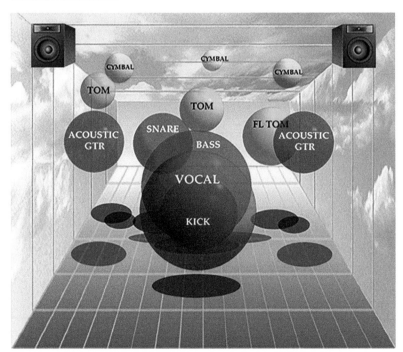

This type of music is typically mixed very clean and clear with hardly any effects. The vocal is normally right out front. Note the presence and complete left and right panning of the acoustic guitars. The snare is set back, and the bass guitar and kick are not too overwhelming.

Visual J. Bluegrass Mix

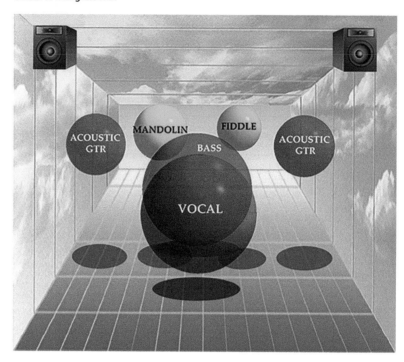

Extremely clear and crispy mix. The volume is relatively even across all of the instruments. When there is a vocal, it is right out front. The bass is set back and is sometimes panned to one side. The lead solos on any instrument are commonly bumped up in volume just a bit.

Visual K. Big Band Mix

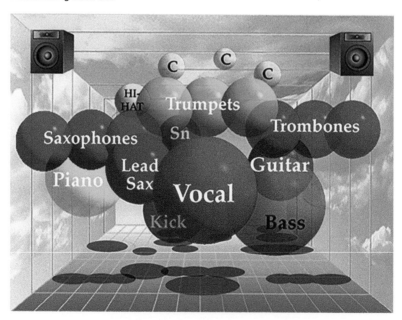

A very clean and clear mix. It's typical for the vocal to be quite loud vocal. The horns are loud compared to rest of the band. The hi-hat is sometimes right out front, while the piano, the guitar, the snare, the bass, and especially the kick are often quite low in volume.

Visual L. Orchestra Mix

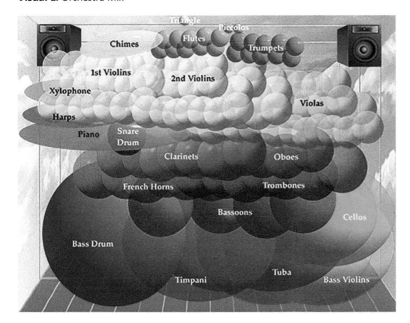

In most classical music, it is important to pan the instruments exactly as the musicians are seated. Not only is this a very strong tradition, but often symphony halls are designed acoustically for such placement. Therefore, you can see that the main bass instruments (bass, cellos, and tuba) are all panned to the right. Note that this visual shows all of the instruments playing at once, which only occasionally happens. The bass drum and tympani rarely play.

Visual M. Balanced Orchestra Mix

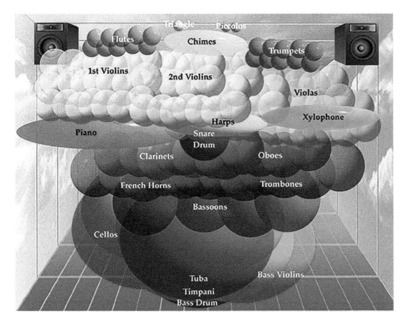

Since classical music is often meant to create a calming, balanced type of feeling, it might be appropriate to create a mix with panning such as this (however, you might go to jail). Just looking at it gives you a more balanced and calm feeling. Also, notice the stereo miking on the piano and xylophone. The low frequencies might even be turned up more to fill out the whole low-end area.

Actual Mixes

Visual N. "She Blinded Me With Science" on *The Golden Age of Wireless* by Thomas Dolby

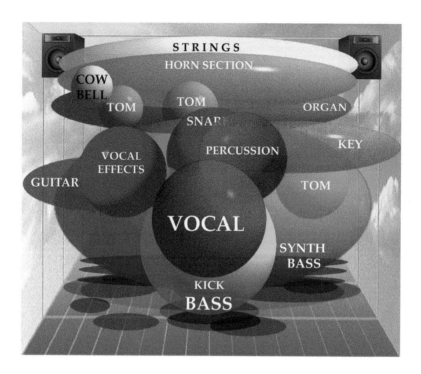

Visual O. The Alarm Clock Section in "Time" on *Dark Side of the Moon* by Pink Floyd

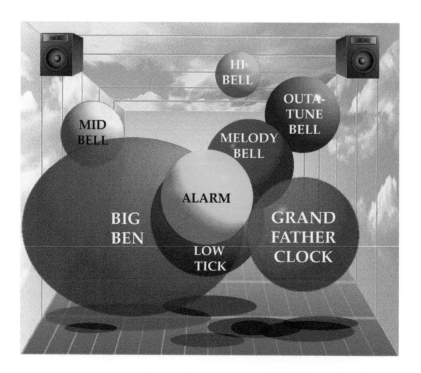

Visual P. "Sledgehammer" on *So* by Peter Gabriel

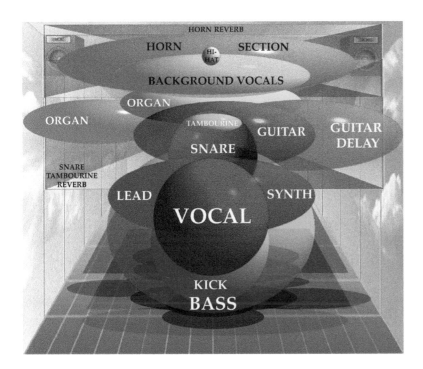

Visual Q. "Babylon Sisters" on *Gaucho* by Steely Dan

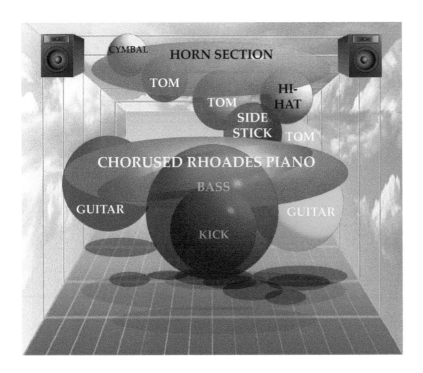

Visual R. Pachelbel's *Canon in D Major*

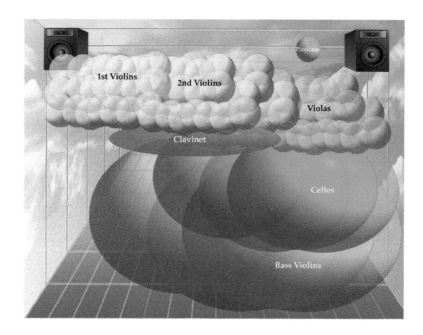

Visual S. "Fallin" on *Songs in A Minor* by Alicia Keys

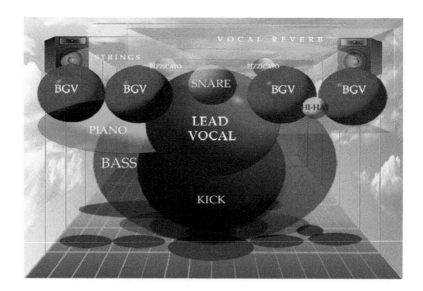

Visual T. "Don't Know Why" on *Come Away with Me* by Norah Jones

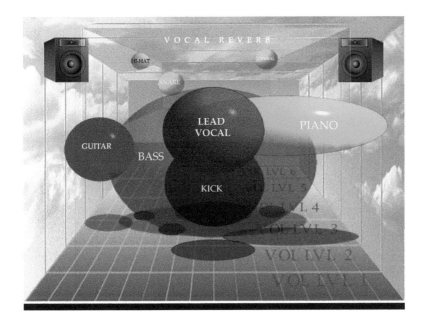

Visual U. "Video" on *Acoustic Soul* by India Arie

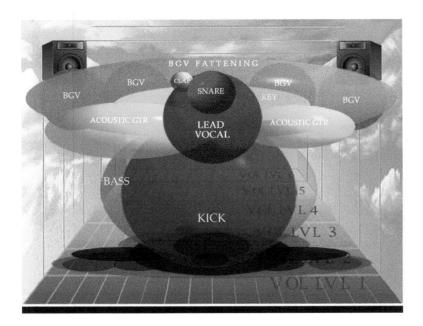

Visual V. "Crash into Me" on *Crash* by Dave Matthews Band

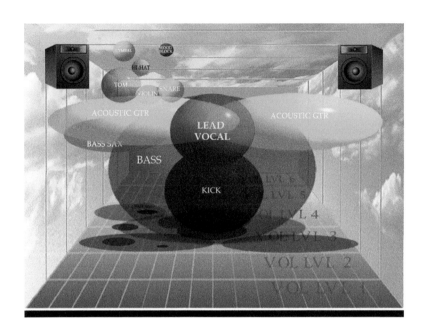

Visual W. "The Way You Move" on *Speakerboxxx/The Love Below* by Outkast

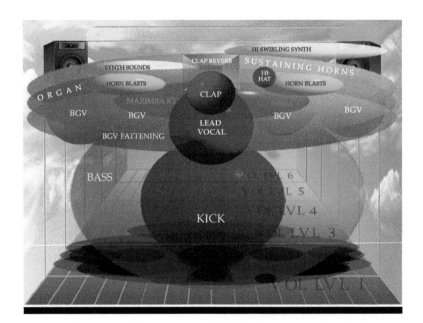

Visual X. "Beautiful" on *Stripped* by Christina Aguilera

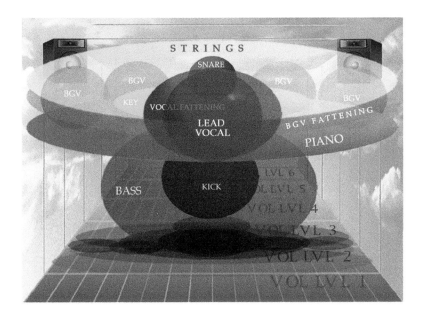

Visual Y. "I'm With You" (Verses) on *Let Go* by Avril Lavigne

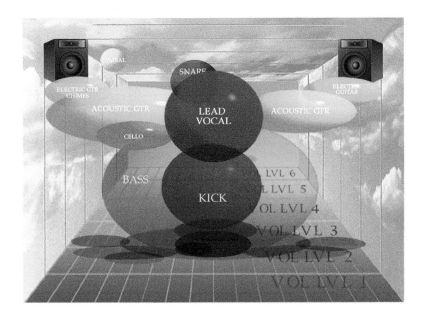

Mix Perspectives

Visual 3. The Thirteen Aspects of a Recorded Piece of Music

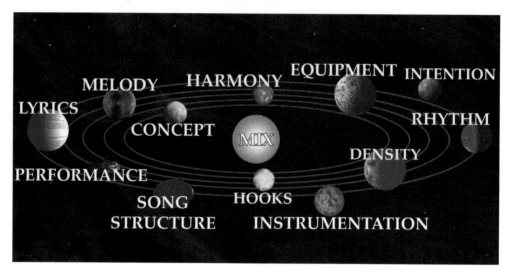

Visual 42. Even Volumes

Visual 43. Uneven Volumes

Visual 44. Symmetrical (Balanced) Mix

Visual 45. Asymmetrical (Lopsided) Mix

Visual 46. Natural EQ

Visual 47. Interesting EQ

Visual 48. Sparse Mix

Visual 49. Full (Wall of Sound) Mix

Favorite Visuals

Visual 1. Sound Imaging of Instruments

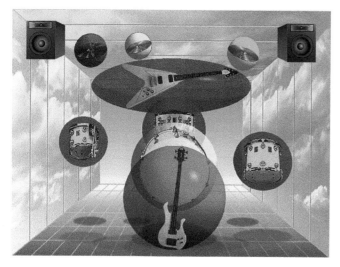

Visual 2. Structuring a Mix

Visual 21. Three Violins with Plenty of Space in Between

Visual 194. Seven Background Vocals Panned to Seven Different Places Combine with Variety of Fattening

Visual 214. Mix with Panning not so Wide Overall

Visual 36. Stereo Reverb on Sound

Visual 50. Virtual Mixer EQ

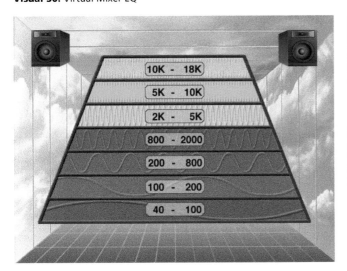

Visual 72. Virtual Mixer Spectrum Analysis

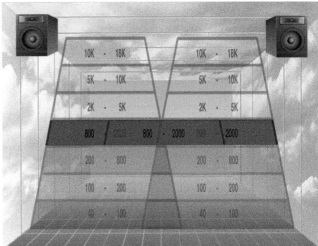

Visual 91. Virtual Mixer Flanging

Visual 101. Reverb: Hundreds of Delays Panned between Speakers

Visual 112. Reverb Filling in Space between Speakers

Visual 132. Six Apparent Volume Levels

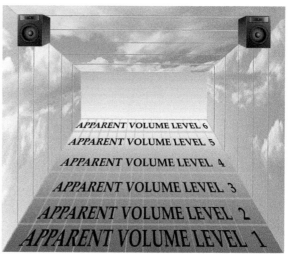

Visual 213. Mix with Extremely Wide Panning Overall

Visual 195. Panning with High End of Piano on Right and Hi-Hat on Left

Visual 196. Simple Mix without Stereo Fattening

Visual 197. Simple Mix with Stereo Fattening

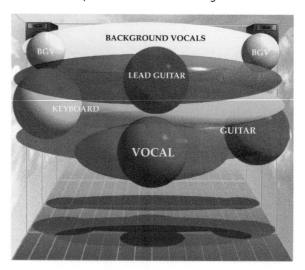

Visual 199. Natural Panning of Drum Kit

Visual 205. Mix Balanced at Each Frequency Range

Visual 206. Unbalanced Mix at Each Frequency Range

Visual 219. Mix with Lots of Different Delays Filling Out the Mix

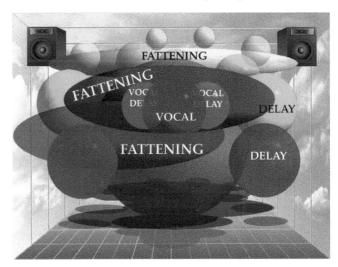

Visual 223. Extremely Busy Mix with no Effects

Visual 224. Extremely Busy Mix with Lots of Effects

Visual 225. Extremely Sparse Mix with Fattening and Reverb

Visual 226. Extremely Sparse Mix with no Fattening and Reverb

Visual 234. Surround Sound Sparse Mix

Visual 237. Surround Sound Mix Throughout the Room

CHAPTER 1

All Aspects of a Recorded Piece of Music

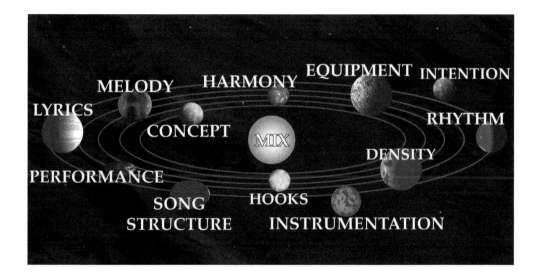

Visual 3.
The Thirteen Aspects of a Recorded Piece of Music

When I did my first album, the mix sounded great, but the band and the song weren't that hot. When people listened to the recording, no one noticed how great the mix was . . . they only commented on how bad the band was. The problem is that most people don't differentiate between the mix and the music. They hear an overall "sound" and make their judgment on that feeling. All I could do was just hope and pray that a great band would walk in the door with a great song.

However, I came to realize that an engineer's job is more than just getting the sounds recorded onto the multitrack and mixing them down. It is also about helping the band or group with their music. Technically, this is the job of a producer. However, engineers commonly function as a producer, particularly when there is no producer in the session (and normally don't get paid for it). In fact, musicians who have worked with professional engineers have come to expect the recording engineer to be able to give them help and feedback on a number of components of the music that you wouldn't normally think of as the responsibility of the engineer.

You wouldn't think the job of an engineer would include working on things like the concept of the song, the melody, rhythm, and harmony. Besides, if the music or band is bad, it isn't the engineer's fault; and making comments in these sensitive areas could be hazardous to your health. However, the more you can help with the music, the better for everyone – the group, yourself, and everyone who listens to it (especially those who don't differentiate between the music and mix). In fact, the engineers who are really good at it often become producers.

Even when there is a producer, he or she often relies heavily on the engineer's opinions. Great producers hire the engineers who are secretly great producers, who have highly refined values in regards to the music.

As you will notice throughout this book, I have been very careful in my choice of words when talking about how often engineers use specific rules or techniques. Therefore, when you see words such as "normally, usually, often, sometimes, occasionally, rarely" – you can be sure that I am trying to gauge, as best I can from my experience, how often these techniques are used. I felt that this was important, because it is my guess that most people who are reading this book are trying to gain perspective on what is common in the industry, especially those who are new to the field.

This chapter covers all of the aspects that go into creating a quality recorded piece of music – those that the producer helps to refine: intention, concept, hooks, melody, harmony, rhythm, lyrics, song structure, arrangement density of arrangement, instrumentation, performance, equipment, and the mix. All of these thirteen aspects contribute to what is perceived as a quality recording and mix. It is important to understand each of these components so that, as an engineer, you can help to refine them whenever possible.

Every one of these aspects of a song should meet at least the minimum requirements of quality. Even though each aspect is only a small part of the overall song, any single flawed aspect could taint the whole song. Even if all of the thirteen aspects are the best they can be, the chances of a song becoming a hit are very slim. If any one of these components is less than excellent, the chances for success go down exponentially. Therefore, it is necessary to critique and refine each of these aspects whenever possible.

The definition of what constitutes "good" and "quality" is extremely subjective and ever changing for each of the thirteen aspects. People often have very strong opinions; therefore, I will be very careful to refrain from making any judgments for you. Whenever possible, I have tried to simply outline ranges of values and preferences so you can ultimately make your own decisions as to which ones feel right to you. Besides, certain values might be appropriate for one song, but not another.

Use these values as a starting point for your lifelong study of people's values in music. Then go out and listen, and base your choices *on what feels right to you* instead of what you think others might like. This is how you become a great engineer. Those who become obsessed with refining the thirteen aspects are called producers. Again, if you decide you want to go deeper, check out my book, *The Art of Producing*, which goes into much more depth on each of the thirteen aspects. Here, I'll cover the common suggestions that recording engineers make in each area. I'll briefly examine each of these components in order to put "the mix" into perspective – as only one component that makes up a quality recorded piece of music.

Aspect #1: Concept or Theme

The concept or theme can be defined as the combination of the other twelve components. It is not the style of music. It is the essence of the song, or the primary message. It is also known as "the mood," "the flow," or "the aura," depending on your perspective. It is normally defined by the lyrics. If a song has no lyrics is often defined by the type of emotion that the song evokes. It is usually a feeling or an idea that is conveyed most consistently and strongly in the most number of aspects of a recorded piece of music.

Songs vary in the consistency or cohesiveness of the concept. In some songs, the concept is quite strong and cohesive, while in other songs it might be non-existent (although the concept could be "no concept"). Normally the more cohesive and consistent the concept, the more powerfully the message comes across. Checkout some of your favorite songs and most likely you will see that the concept of the song is supported by every musical component, as well as the mix.

Admittedly, there are some songs that make no sense at all that are incredibly cool, fun, or effective. This is because other of the thirteen aspects are just over the top well done.

As an engineer, it is important to first try and identify what the concept of the song is about. If it is unclear, you might even ask the musicians.

Then, as you get to know the song, keep an ear out for anything that is being played that doesn't fit or support the concept. Often a part might need to be taken out or played with a more appropriate feel. For example, you point it out if a guitar part doesn't seem to fit the "feel" that the rest of the band has already established. Or, maybe you might mention that the effect that is being used by someone on their instrument doesn't seem to fit the "mood" of the song. Perhaps someone wants to use a sound or an effect that they heard in another song, but it isn't appropriate for the current song. It is the engineer's responsibility to point out these inconsistencies (with kid gloves, of course).

Also, notice how all the other aspects relate to and contribute to the cohesiveness of the whole song.

You might also suggest ideas that help make the concept stronger and more cohesive. Such detailed analysis can sometimes provide inspiration and lead to the creation of new ideas. Perhaps you might suggest repeating an especially relevant section, adding another section that enhances the concept even more. You might even work with musical parts or lyrics to make the concept more clear and give it a more profound meaning.

On the other hand, it is important to be careful to detect creative ideas of genius that could easily be misconstrued as inappropriate. Sometimes really cool ideas just may not be appropriate for a particular song and concept.

Once you know what the concept of a song is about then you can use that as your guiding light. You can then go through all of the other thirteen aspects and check to see if they are supporting the concept.

> Whenever making comments about anyone's music or contribution to a song, it is helpful to end the statement with, "I don't know . . . what do you think?" This is not only considerate, it opens the doorway for the musician to give their opinion. It can also help desensitize any overly sensitive egos.
>
> How many producers does it take to screw in a light bulb?
> I don't know . . . what do you think?

Aspect #2: Intention

Intention involves holding a certain energy or emotion throughout the recording and mixing process *with 100% focus!*

I would assume for most of you reading this (even those who have been around awhile in the business) that intention might strike you as unusual or curious. However, it has become the most important aspect in the music that I have been doing over the last 15 years.

The intention of a song is very similar to the concept of the song, but much simpler. For example, maybe the concept of a song is about a love relationship, however, the intention might be simply, "love." Then you would hold the energy of love throughout the song. Not only would you as the recording engineer hold that energy, but you would help the band to hold that energy throughout the whole song.

Although this may sound a bit "new agey," there is good scientific evidence to show that intention when held with 100% focus gets embedded in the music and carried through to the listener.

But even more importantly, it affects the way the musicians play. For example, if a band is enveloped in the energy of love while recording (as well as the engineer), then they will certainly play better. It often helps musicians to get "out of their head" and focus more on the energy of the song. Besides the key aspect of great music is music that powerfully evokes an emotion. The more people are in a particular emotion, the more it is conveyed to the listener.

This technique may not work so well if the song is about hate (or maybe it would . . . God forbid).

Meanwhile, if the intention is about something as powerful as "universal love," or "connecting to source energy," then a whole other level of creative energy often ensues.

Again, this also means holding that energy when performing the mix.

Ultimately, when a band even holds an intention with 100% focus when writing the song, it is even better.

I have a song called "Unconditional Love," and I was very careful to first invoke the energy of unconditional love before I started writing the song. I then noticed my breath as I meditated. I then set the metronome in the Digital Audio Workstation program to the same tempo as my breath. I was careful to stay in that energy while writing the song and recording each part. When I had other musicians come in and play on the song, we would first do a little meditation and bring in the energy of unconditional love. Even while mixing, we all held that energy. That song is now packed with positive intention and is much more likely to invoke the energy of unconditional love when listened to.

Try it and see what you think. It may not be appropriate for every band or group.

Aspect #3: Melody

The first question to ask is whether the melody seems to be supporting the concept. It should not take away from or distract you from the concept or intention.

The next most important thing is to see if the melodies are too busy or too simple. Sometimes they might be simplified. If too simple, you might suggest making them a little more interesting, or adding some interesting effects to them to spice them up – and avoid yawns.

A person's melody line can be quite personal. Therefore, commenting on someone's melody line can be dangerous. People can often be attached to the melody they came up with. Sometimes, it just might be what the band or group really wants. Therefore, a little tact can be in order.

In order to avoid copyright problems, you would want to comment if you find that the melody line is exactly the same as another song.

As an engineer, you would probably point out a bad note in a melody, but it might turn out to be intentional. You might also come across a case where the band is improvising around the melody in the choruses of a song (especially in jazz). You could mention that it might be a good idea for them to go ahead and sing or play the melody line straight in the first chorus in order to establish the melody.

Aspect #4: Rhythm

Those of you who know a lot about the complexities of rhythms, or simply have beats going around inside your head throughout the day, might make some suggestions if you feel it is appropriate. There are entire worlds of rhythm that are taught in music theory classes.

However, even if you know nothing about rhythms, there is still one thing you could critique: Is the rhythm too busy or too simple? If you are falling asleep, it could probably use some spicing up. If you can't keep up, the rhythm might need to be weeded out a bit.

Also be on the lookout for when one rhythm part is not working with another. Perhaps the guitar part is stepping on the keyboard part. If so, say something. In fact, if any part is bugging you, don't be afraid to bring it up. If they don't agree, let it go (unless it is really bugging you).

It is also a good idea to check out the variation in rhythm parts from section to section in the song. For example, you might suggest that the drum pattern be changed a bit for the lead break or bridge. A change in the guitar part might be appropriate for the choruses. Perhaps the way that the rhythm parts change from section to section doesn't work for you or is distracting. If so, you might say, "Hey, is that really the way you want it?"

It is also good to notice whether the tempo of the song feels right to you. If the vocals sound rushed or too laid back, you might say something. Sometimes you might just sense that it would feel better a little slower or faster. Bring it up and see what they think.

Aspect #5: Harmony

First, if you can arrange the harmony parts for the band, they will often be amazed, that is, if they can sing them. Even if you don't know anything about chord structures, inversions, or voicing, most people can tell if it just doesn't sound right somehow. If so, point it out.

Besides critiquing the actual notes in the harmony parts, you can also make suggestions about the number of parts and their ultimate placement in the mix. A band may not be aware of all the different ways that background vocals can be recorded.

You could record a three-part harmony on one microphone and place them in the left speaker in the mix, then record the same three parts on another track and place them in the right speaker, creating a full stereo spread of harmonies. You could also record three people in stereo with two mics. By putting one mic in the left speaker and the other in the right speaker, you will hear the three parts left, center, and right between the speakers. Then, record the same three parts again with two mics and place them so that you have two parts of each left, center, and right. You could also record a three-part harmony sixteen times on sixteen different tracks so that you have forty-eight vocals. Then "bounce" (record) the sixteen parts onto two open tracks in stereo. Once you have mixed the sixteen tracks down to only two tracks, you can then erase the original sixteen tracks and reuse them. You end up with forty-eight vocals on two tracks in full stereo for that huge Mormon Tabernacle Choir effect.

Many bands don't realize all the possibilities for recording background vocals, so they often appreciate it when the engineer suggests possibilities that might be appropriate. I will commonly have the band at least record the background vocals twice. Then place them completely left and right, so they feel balanced between the speakers.

Aspect #6: Lyrics

I used to make comments about bad grammar in the lyrics. For health reasons, I now refrain. Just think of all the hit songs that have bad grammar, stupid lyrics, or lyrics that just make no sense at all. Therefore, you might use a little bit of discernment and restraint when making comments about someone's lyrics, unless you know the people well.

If you add one word to a song, you then own part of that song by law. If the lyric is an important part of the song, it can be a large percentage of the song. As you can imagine, this makes some songwriters quite wary of taking any suggestions. Therefore, because of copyright laws, an engineer should be extremely careful when making suggestions in this area. Instead of coming up with some new lyrics, it is always better to try to get the songwriter to come up with new ideas themselves.

One of the most important things to watch out for is that the lyrics are rhythmically correct. In most styles of music (especially rap), it is critical that the lyrics fit the music rhythmically. If you hear lyrics that have too many or too few syllables so that it doesn't seem to flow, you might say something. Ask the band if they are happy with the way the lyrics work with the music rhythmically. If they're not, try to get them to come up with something else. Going out on a limb to help refine any weak lyrics could mean the difference between a hit or just an overall great song.

Aspect #7: Density of Arrangement

The density of the arrangement is defined as the number of sounds in the song at any single moment, including how many sounds are in each frequency range.

The main aspects to evaluate in an arrangement are the sparseness or fullness. If the band is obviously trying to create as full a mix as possible, you might make some suggestions to help. Adding more sounds or notes is the best way to fill out a mix. Therefore, you might suggest double-tracking (recording the exact same part twice) or even triple-tracking. You could recommend doubling a part with a different instrument or even suggest that someone play a busier part. You can also mention that adding time-based effects, such as delays, flanging, or reverb, will help to fill out the arrangement. Recording a part in stereo with multiple microphones will also add to the fullness of the song.

However, a more common problem with arrangements is that they are too full and need weeding out. There are some bands that would record forty-eight tracks if available, just because they are there and they can! Even worse, when it comes to the mix, they want all forty-eight tracks in the mix because they have become attached to their parts. After all, they spent so much time recording them. Even if there isn't enough room between the speakers for all of the sounds, they want it all in there because they played them.

It is often helpful to try to weed out the arrangement so you can hear each instrument better. An engineer will often suggest turning off (muting) certain tracks in particular sections of the song. Dropping out parts like this can make certain sections of the song seem more personal, contributes to the overall clarity, and helps create a sense of dynamics that breakdown and then build back up.

Bands often don't think about dropping out instruments or sounds from the mix because they are used to playing live onstage. It might never occur to them to actually stop playing in certain parts of the song: "What do you mean? Stop playing?!" Often, simply demonstrating what it would sound like when a part is dropped out will convince the band. Of course, it is especially common when mixing hip hop or dance music to turn various tracks on and off throughout the mix.

On a more detailed level, the engineer might point out when too much is going on in a specific frequency range. In this case, one sound can hide another (masking). You might suggest playing a part at a different octave or in a different inversion.

Aspect #8: Instrumentation

As an engineer, you're responsible for making sure that each instrument sounds good, regardless of whether you or the band chose the instrument. This means two things: (1) there is nothing wrong with the instrument that is causing a problem with the sound and (2) you have a good quality instrument that is putting out a "good quality" sound. A "good quality" sound could mean interesting, complex, unique, beautiful, intense, or in some instances, annoying.

Problem Sounds

Problems include extra noises and buzzes, or just odd tonality. If there is something wrong with an instrument sound, you can only do so much to fix it in the mix, no matter how much you process or effect it. Therefore, it is important to recognize sounds with problems in the first place, so you can replace them. If you can't replace them, point them out so that the band realizes their instrument sound was bad, not your mix. Hopefully, this will motivate the band to get the instrument fixed for future recording sessions.

On drums, make sure all the heads are new. Let the band know that there is nothing in the control room that can fix a drum head that's held on with duct tape. Also make sure that there is nothing wrong with the guitar or the amplifier. Each guitar should be set up so that the intonation is correct. This means that both the nut and the bridge (which hold the strings up at each end), and the frets are in the precise spot to make every note be perfectly in tune. If this is not the case, then even after tuning the guitar with a tuner, certain notes will still be out of tune. Also make sure there is no ground buzz in the guitar amp.

Quality Sounds

Even though it is technically the job of a producer, professional recording engineers should know the difference between different brands and types of instruments intimately. As an engineer, it is really helpful (and often appreciated) if you can suggest and obtain (borrow or rent if necessary) higher quality instrument – or suggest a uniquely appropriate sound inside of a sound module or synthesizer.

If you have a drum kit that doesn't sound so great, see if you can rent another set. There is no reason for a guitar player to use the one guitar they have for the entire album. Beg, borrow, or steal a selection of guitars for the project. The album will normally sound much better with a variety of guitar textures.

Make sure all the amps sound good. Again, there is no reason for you to have only one amp sound on an album. It is especially effective to use a Y cable so one guitar can feed two different amps. Recording the two amps on separate tracks gives you a variety of sound combinations, creating a totally unique sound.

You can study different instruments in books, but to really get to know them, you need to hear them. One method is to hang out at your local music store. Another way is to pay extremely close attention to instrument sounds that you come across in concerts and in the studio. Of course, learning to play various instruments is helpful. If you come across an instrument where you are unfamiliar with what a quality sound is, ask the musician what they think of the sound. They will often have comments which can help you to learn about what makes a great sound in a particular style of instrument.

You should also become familiar with all of the sounds within each synthesizer in the studio you are working in so that the band won't have to spend 2 hours going through all 5,000 presets. You can easily direct them to the presets with the types of sounds they are looking for. An engineer will commonly suggest unique sounds to incorporate into a song. There are a huge number of totally unique and bizarre sounds that could be used. There is a world of different ethnic percussion instruments available these days. You might even suggest designing a new sound altogether with a synthesizer or computer. You might also think about sampling some unusual natural sounds and using them as instrument sounds. When placed low in the mix, some very unusual sounds can actually fit in quite well, even in the most normal type of song.

Aspect #9: Song Structure

The song structure refers to the order and length of the song sections (intro, verse, chorus, lead break, bridge, and vamp). As an engineer, you wouldn't normally say very much about the structure of someone's song, unless you knew the songwriter well. However, if the structure somehow bothers you, or if you have an idea to make it better, you might ask the band what they think.

For instance, you might point out that a 5-minute lead break is a bit long, and people might not be patient enough to listen to an introduction that is longer than 10 seconds. You might suggest that they do two different versions or that you edit the final mix to make a shorter version.

Aspect #10: Performance

The recording engineer – whether there is a producer on the project or not – is most often responsible for critiquing and refining a performance. There are five main aspects of performance that a recording engineer may be involved in: pitch, timing, technique, dynamics, and greatness (the goosebump factor).

Pitch

Normally, the recording engineer is ultimately responsible for all instruments being in tune and every note of a performance being in pitch.

There are two levels of pitch perception. *Perfect pitch* is when you can recognize the exact note or frequency of a sound. Some people can tell you the frequency (such as 440 Hz) when they hear a note. This skill, although great to possess, is fairly rare.

Relative pitch, the ability to tell if a sound is in tune with previous sounds in a song, is much more common and is extremely important. It is critical for a recording engineer to develop good relative pitch perception. Some people are born with it. However, it is a skill that can be learned. There are also some very good computer programs and study courses on tape that teach this skill.

I believe that most people have very good pitch, if they are able to hear the sound for a long enough time. It is more of an issue of getting quick at listening. Ask yourself if the sound is in pitch at *every single moment*. It all comes down to simply concentrating on finer and finer moments. You need to be able to hear the relative pitch of each note in the beginning, middle, and end of each note. As your concentration gets better, you can hear the pitch of every single note in a string of notes at a fast tempo. This amount of concentration becomes easier with practice.

The hard part is getting to the point where you can remember which note is out of tune in an entire riff. It is great if you can also tell if a note is flat or sharp, but it is not absolutely necessary. It is enough to know simply if a note is out of pitch and which one it is.

Learning to be able to tell if a sound is in tune or not requires the discipline of focusing on every single moment. The good news is that over time, it gets inside your body. Now whenever I hear a sound that is out of tune (even if it is in the background), it jumps out and grabs me.

Timing

Another important skill for a recording engineer to master is being able to tell if the timing is stable. Some people are born with perfect timing perception. However, most of us have to listen closely. Some people pat their leg. Others keep time with one finger in the air. Some just sit with their backs straight up and tilt their head in a funny way. Regardless of the technique, it takes serious concentration to hear variations in tempo.

Some people are fanatics about stable tempo and go out of their way to get tempos to be as stable as possible. Therefore, it is critical to find out if the band's values lie in this area so that you can give the necessary added attention and focus to the timing. If a band doesn't care, work with them to get them to focus on it more.

You might also suggest having the drummer play to a click track. However, it often takes months of practice before a drummer gets comfortable enough with a click track so they are not always chasing it. It takes even longer for a drummer to get where they can play a "feel" on top of the click track. If they haven't practiced, have them rehearse for awhile with a click track. Then when you go to record, take the click track out of their headphones. They will now be more focused on timing.

Regardless, always listen closely for whether the drummer is consistently on the beat. Of course, sometimes a drummer will intentionally play on top of the beat, or behind the beat to create a certain feel. Often, drummers will speed up when playing a tom-tom roll, or when nearing the end of the song. Also, watch out for someone who is coming in too early or too late on a vocal part. Same with a rhythm guitar part, listen closely to their consistency in timing.

Do what you can, but there is only so much you can do before they get irritated. Always try to calculate how far you can push, or lead them.

Technique

There are specific techniques that musicians must learn to play an instrument correctly, technically. These will vary depending on the style of music being played. Any tips or techniques you can offer to a musician can only help. Of course, you can't be expected to know the right thing to say to a musician for each and every instrument; but the more you work in the business, the more tricks you pick up and the more you can help out.

For example, there are specific techniques for playing each of the drums in a drum set. The kick drum should be "popped" with the foot. For some styles of music, it is best to really whack the snare drum. For guitar players, there are many little things to watch out for, such as not causing any string buzz from hitting the frets or not causing the strings to squeak as you move up and down the neck. All that is necessary is to point out the issues when they happen.

There is a wide range of comments that you can make to help singers (as well as a wide range of comments that don't help). Suggestions such as, "sing out," or "project more," can be helpful if given at the right moment and with sensitivity. Often it helps to get the singer to focus more on using their diaphragm. Some people will even have a vocal coach come into the studio during the recording session to help out.

It is especially important to pay close attention when an experienced producer or professional musician makes a comment or suggestion that works. Remember each of these tips and, after awhile, you will learn an entire range of techniques that you can use to help musicians play better.

Dynamics

There are two main types of dynamics that you can critique and help refine. First, it is a good idea to keep an eye on simple changes in volume dynamics in a performance. You might find them to be too dynamic when they vary too much. Or you might find them to be too stable in volume so that it sounds like a synthesizer or drum machine. It is important to make sure that the volume dynamics fluctuate in a way that is musical or appropriate for the song.

The second dynamic to critique is the level of emotional intensity at every moment in the song. Just as with volume dynamics, you might find them to vary too much, be too boring, or be inappropriate. For example, singers occasionally sing out too intensely at the beginning of a song when perhaps they should be saving it for the end of the song. On the other hand, maybe they need to put more emotion and feeling into the performance right up front.

Checking out the performance dynamics at each moment in the song can help you fashion it in exactly the way you want.

Greatness

This is the "goosebump" factor. You should never let a performance go . . . until it turns you on. There is a wide range of values that people hold. Common values include sincerity, heartfelt feelings, and emotions. Most likely, you are in this business because you know what you like. At the very least, don't let a performance stand that you don't like. If you make sure that every single performance is incredible, at least in your eyes, chances are that the overall performance of the song will be great.

Over the years, I have been leading discussions in class about exactly what "greatness" is for different people. People use different words to describe it. I often say it is "happenin." I used to say "cool" too much. Now my favorite word is "WOW!" However, "tight" seems to be the deal in a lot of hip hop and "bad" is still good. Think about the words you use, and whenever you use the word, stop and take a closer look at exactly what it is that is causing you to use it. This way you can begin to define, learn, and remember everything that is incredible.

I have also done a survey of my classes over the years and have found that 99% in this business do get goosebumps or chills when they listen to some music that they really like. I was surprised at this. I would have guessed that only maybe 50–70% get chills. So chills are rampant in our society. Even more interesting is the wide range of chills that people get. Sometimes, I get them throughout my head. But when something is really incredible, I feel them through my whole body down into my toes. Therefore, ideally, you are looking to get chills from every performance you work on.

> Perfection has no limits! Once you obtain perfection you have then advanced to the next level, where you can see how much better it could be.

There are a number of factors that contribute to the decision of how much time you spend trying to get a great performance. It is the engineer's responsibility to gauge the amount of time spent on refining a particular performance, particularly when working on a demo project instead of a CD. Regardless of the circumstances, everyone wants a basic level of quality. However, after obtaining this basic level of perfection, there is only so much you can do to get an excellent performance. This will be dependent on the following factors.

Budget

If the band can't afford the time to perfect a performance, there is nothing you can do unless you are rich or own the studio and are extremely generous. If the band is trying to do a ten-song demo for $100, you just might have to move the session along.

Deadlines

A deadline, such as a meeting with a record company, an appointment to have a project mastered or pressed, or even Christmas (especially when a project is rushed to be ready for holiday sales), is one of the primary destroyers of project quality. It can often help to point out to the band how detrimental deadlines can be. However, sometimes they cannot be avoided, so if a group has a limited amount of time, an engineer might have to accept a performance that is less than perfect.

Purpose of a Project

Obviously, if a project is destined to be a CD, much more refinement is in order. Vinyl is final, and every album is a part of your reputation. If the project is being done as a demo, then the engineer might let less-than-perfect performances pass as acceptable. Generally, an engineer will try to obtain the best performances possible on drums because of the amount of setup time involved. If the demo is accepted by a record company, the drums could then be kept as basic tracks for the album.

Expertise of Musicians

The quality of musicianship makes a big difference in the amount of time it takes to get an acceptable performance. You would think that the worse the players are, the longer it would take. But this is often not the case. Many times, great musicians take even longer because they know how good they can be. If you have a musician who is not good enough, you might need to suggest hiring professional musicians (if you have a bulletproof vest or enough tact). One good idea is to present this idea, then tell the band that if they like their own playing better than the professionals, you'll pay for it. I've never had to pay for it yet.

Apparent Musical Values

Different people hold different values for their music. For example, a punk band might focus on energy instead of perfect tuning. An R&B band might care about the spatiality of the sound. A rap group may be mostly concerned about the "boom." A jazz combo might emphasize the interaction between the players. Often, these values will determine whether a performance is acceptable or not. It is often fruitless to spend too much time on an aspect that the band couldn't care less about. However, there are certain basic levels of expertise that you should demand. On the other hand, it is critical to pay extremely close attention to the aspects that the band obviously values the most.

Determination

The amount of determination that a band brings to a project affects the time spent working on a part and the quality of the final project. Often band members don't realize how much work it takes to get a performance perfect or great. Musicians can easily get frustrated or fatigued to the point where they say, "Good enough." You should always try to inspire everyone to work harder and longer until it is as good as possible, but you can only push musicians so far before they become irritable. It might help to simply point out that it is normal for it to take a long time to get things right and that professional musicians often take days to get a performance perfected. This can help to inspire people to push themselves to be great.

On the other hand, some musicians are so determined to get a performance perfect that they never stop. In the beginning, these people can make you nuts, but you'll soon realize that with this kind of perfectionist, you will end up with an incredible performance. Subsequently, when people listen to the project they will say, "Wow, you recorded that?" Therefore, you come to appreciate the obsessive ones.

Aspect #11: Quality of the Equipment and the Recording

The quality of the equipment refers to the recording equipment, as opposed to the instruments (which were covered under the "Instrumentation" section). The engineer should make sure that all of the equipment is of the best quality possible and, even more importantly, that it is in good working order. These days, this means good quality microphones, mic preamps, processing, and effects.

The quality of the recording includes things like getting good recording levels (not too low or too hot), good miking techniques, and no distortion or excessive noise. Obviously, these are the recording engineer's responsibility.

Aspect #12: Hooks

Hooks are things in the song that are memorable. The most common hooks are lyrical hooks. However, any of the thirteen aspects might stick in your head. Things that stay in your head after listening are really important – because most people won't go out and buy a song, or play it again, if they can't remember anything about the song.

Hooks range from those that are so stupid you can't get them out of your head, to those that touch you very deeply. The best hooks are those that touch masses of people deeply.

As you will notice, when you start checking it out . . . the number one thing that makes a song a hit (or great) is the quality and number of hooks in the song.

As a producer/engineer, it is important to look out for hooks. Again, you could find that each of the following has something memorable (I've also included examples):

- Concept – one that touches you deeply
- Intention – so strong you feel it in your Soul
- Melody – you can't stopping humming it afterward
- Rhythm – your body is still moving afterward
- Harmony – so beautiful
- Lyrics – they keep going around in your head
- Density – the arrangement builds with more and more instruments mellows with less and less.
- Instrumentation – just really cool instruments or sounds
- Song structure – unique order of sections of the song
- Performance – chills!
- Quality of recording and equipment – well recorded with high quality equipment
- Mix – appropriate for the song with interesting dynamics

Since hooks are so important for the overall greatness of a song, I will often recommend using a hook in other parts of the song. If you have something great, try and sprinkle it around as much as possible – without over doing it!

For example, you might use a great melodic hook for the intro of the song. You might even put it low in the background in the intro to just "hint" at it coming up later. You could even have a different instrument, play it.

If you get inspired, you might also introduce new hooks. I often like to add an interesting sound of melodic riff in a "tag" section (a little lead bread section after the chorus that has no lead instrument actually playing).

Aspect #13: The Mix

The mix is only one part of everything that goes into creating a great overall recording; however, it is one of the most powerful aspects because the mix can be utilized to hide weaknesses in other areas and create its own magic.

The rest of this book is about the mix.

Homework for the Rest of Your Life – Due (Do) Every Day

Quality is defined in different ways by different people, so it can take a while to learn all the ways in which songs can be refined. When it comes to values, the only one that's really bad is "no values at all." To develop your own values, start focusing on each of the thirteen aspects whenever listening to music. As you check out the details of each of these components in songs you listen to, you will develop a range of values. Even if you know nothing about music theory, you can still learn how to make careful suggestions about each of these components. Then you will have more to offer in the recording session.

Whenever you have time to listen closely to a song (I often do it while driving), critique each one of the thirteen aspects. Try and define what the engineer or producer did for each aspect. Then ask yourself, "Do I like this or not?" At first, the answer to this question might be, "Don't know." However, if you simply start paying close attention to each component, you will naturally develop your own perspective on what you like and what is "good."

Meanwhile, many of you have already developed some pretty detailed ideas as to what quality means for each aspect. In fact, it seems some people are born with it. From years of teaching, I have found that many people already have very specific and highly developed values, but they often have never articulated them. So the trick is to define them and put the values into words. Not only does this help you to remember the values, you will be more confident to share them in recording sessions, whenever appropriate.

Another good exercise is, whenever you hear a song that you really like, ask yourself why you like it – Which one of the thirteen aspects is it that is making you like the song? It might be a combination. This way, you can begin to pin down your own values for each of the thirteen aspects. Inevitably, your values will not only start to shift, they will also become deeper and more refined.

It is good to get into the habit of critiquing each of the thirteen aspects on every song that you have the time to listen to in detail. The truth is, this is what professional engineers and producers do all the time. At first, it is tedious. Later it becomes second nature. Ultimately, it helps enrich your listening pleasure because you are able to get more depth out of the music you listen to. But most importantly, it expands and deepens your range of values so you have something to offer in a recording session.

CHAPTER 2

Visual Representations of "Imaging"

Section A: Physical Sound Waves Versus the Imagined Placement of Sounds between the Speakers

We relate to sound in two ways: we feel (and hear) the physical sound waves that come out of the speakers, and we imagine the apparent placement of sounds between the speakers.

Physical Sound Waves

Whether in the control room or living room, sound first comes out of the speakers in sound waves and travels through every molecule in the room, hitting all parts of your body, and goes into your ears. Just as waves travel on water, sound waves travel through the air. When the speaker pushes out, it creates compressed air (denser air with a higher air pressure) in front of the speakers. This compressed air corresponds to the crest of a wave in water. When the speaker pulls back in, the sound doesn't return. It creates "spaced out" air (rarefied air). As we all know, when you are in the tub and you push the water and pull your hand back, the water doesn't come back. Instead, a trough is created. In the air, this trough corresponds to spaced out air. Therefore, sound travels in waves consisting of alternating compressed and rarefied air. This is one way that we perceive sound.

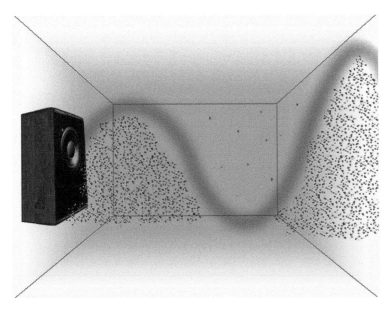

Visual 4.
Sound Waves:
Traveling Compressed
and Spaced Out Air

"Imaging"

The second way we perceive sound is by imagining sounds between the speakers. The apparent placement of sounds between the speakers is called "imaging" because it is a figment of our imagination. So you see, we're not talking about reality here. When we imagine a sound, like a vocal, to be between the speakers, there is, in actuality, no sound there. The same sound is coming out of both speakers, traveling throughout the room, and we just imagine the sound to be between the speakers.

The same thing happens when you listen through headphones: when you hear a sound in the middle of your head . . .

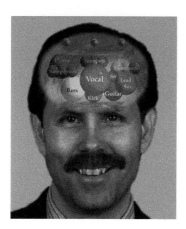

Visual 5.
Imaging in Head

. . . there's no sound there. Your brain's there!

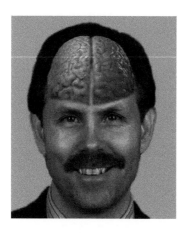

Visual 6.
Brain in Head

With no imagination process, such as when you are asleep, there's no imaging. If you aren't paying attention to a mix or if you are off to the side of the speakers, imaging does not exist. If a tree falls in the forest and you aren't there, there is not imaging. On the other hand, physical sound waves still hit your body when you are asleep. Even if you aren't paying attention, sound waves are still physically vibrating every cell in your body. You feel sound waves even if you aren't listening. When a tree falls in the forest, the sound waves still physically affect every other plant and animal in the area.

Imaging requires active imagination to exist. Sound waves do not.

Although professional engineers utilize both modes of perception to gain as much information about the mix as possible, they are often more concerned with the dynamics that exist in this imaginary world of imaging.

A wide range of dynamics are created by different placements of sounds between the speakers, and these dynamics are utilized to create all the various styles of mixes that fit all types of music and songs.

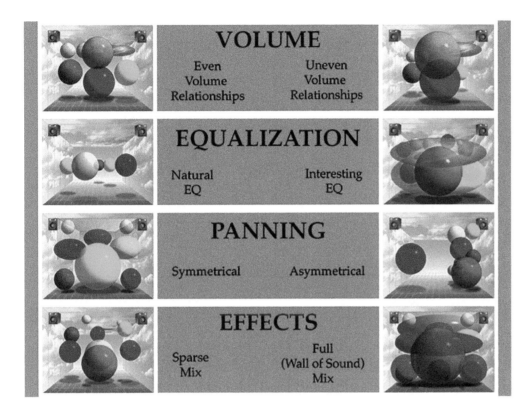

Visual 7.
Dynamics Created with Studio Equipment

Section B: The Space between the Speakers

Mapping Volume, Frequency, and Panning Visually

In order to explain different styles of mixes, let's map out how each piece of equipment affects imaging, the apparent placement of sounds between the speakers. There are three basic parameters of sound corresponding to the X, Y, and Z visual axes.

Visual 8.
Sound to Visuals:
X, Y, Z Axes

Panning as Left to Right

Panning, the left/right placement of sounds between the speakers is naturally shown as a left to right placement visually.

Visual 9.
Panning: Left to Right
Placement

Volume as Front to Back

Sounds that are closer to us are louder and distant sounds are softer, therefore, the volume of a sound in the mix can be mapped out as front to back placement.

Visual 10.
Volume: Front to Back
Placement

As you have probably noticed in mixes, some sounds are right out front (normally vocals and lead instruments), while other instruments, like strings and background vocals, are often in the background (consequently, the term *background vocals*). If you want a sound out front in a mix, the number one thing to do is to raise the fader on the mixing board. Lowering the volume will, of course, put the sound in the background.

Although volume is the number one function of front-to-back placement, other studio equipment can be used to make sounds seem more out front. Boosting an equalizer at any frequency range will normally make it more out front because the overall sound *is* louder. Boosting midrange frequencies will accentuate the presence of a sound more than others, making it seem even more in your face. Compressor/limiters can also be used to make sounds more out front. They do this by stabilizing the sound so it doesn't bounce up and down so much in volume. When a sound is more stable, our minds can focus on it more clearly, making the sound more present. Short delays less than 30 ms, called *fattening*, will also make a sound more present. Also, certain harmonic structures of sounds will stick out more than others. For example, a chain saw will cut through a mix, much more than a flute. Time-based effects, such as chorus and flanging, and longer delay times tend to make a sound less present simply because it is obscured by a second delayed sound. All of these effects are discussed further in Chapter 4, "Functions of Studio Equipment and Visuals of all Parameters."

> NOTE: In reality, you need other cues, such as delays and reverb, to help gauge the distance a sound is from you. This is because you could have a loud sound that is really far away, or a soft sound that is really close. However, if you have a sound that is playing the same volume, and you raise or lower the fader level, technically you don't know if the sound is getting more distant or just getting softer in volume.
>
> If you happen to be hanging out in an anechoic test chamber (a room that absorbs all sound so there are absolutely no reflections off the walls), you can't tell the distance of a sound by volume alone. However, for the purposes of creating a framework as a tool to explain all the structures of mixes possible, it works well to show volume as front to back. After all, the louder the sound, the more out front it will appear in the world of imaging and mixing.

Pitch as Up and Down

There is an interesting illusion that occurs with high and low frequencies in the world of imaging – highs are higher and lows are lower. Instruments such as bells, cymbals, and high strings always seem to be much higher between the speakers than instruments, such as bass guitars, kick drums, and rap booms. Check it out on your own system. Play a song and listen to where high- and low-frequency sounds seem to be between the speakers. Height is especially noticeable in a car.

Visual 11.
Frequency: Low to High
Placement

Visual 12.
Frequencies in Us

There are a number of reasons why this illusion exists. First of all, low frequencies come through the floor to your feet; high frequencies don't. A piccolo will never rumble the floor. In fact, professional studios are calibrated to exactly how many low frequencies travel along the floor to your feet. (This is why some engineers like to work barefoot!)

Another reason that explains why highs are high is the fact that our bodies have a large resonant chamber, the chest cavity, below a smaller resonant chamber, our head. Voice instructors teach you to use these resonant chambers to accentuate different frequency ranges. If you want to bring out the lows, resonate the stomach.

On a more esoteric level, there are energy centers in the body, called *chakras*, that respond to different frequencies. These frequencies are specifically mapped out from low to high, from the base of the spine to the top of the head.

I believe the placement of these energy centers in the body is most likely the main reason we perceive high and low frequencies as high and low. But regardless of why it happens, the truth is that high frequencies do seem to appear higher between the speakers than low frequencies. This is also probably why they call high frequencies "high," and low frequencies "low." Therefore, I put the high frequencies up high and the low frequencies down low in all the visuals.

Visual 13.
Song with Highs and Lows Highlighted

You can raise or lower a sound by changing the pitch with pitch shifters, harmony processors, and aural exciters or by having a musician play their instrument in a higher octave or chord inversion. As I'll discuss later, this becomes important when one sound masks, or hides, another at a particular frequency range. Spreading sounds evenly over the frequency spectrum can help make your mix much cleaner and clearer in the first place. Since equalization controls the volume of frequencies, with an EQ you can move a sound up and down . . . at least a little bit. No matter how much bass you add to a piccolo, you will never be able to get it to rumble the floor, and adding treble to a bass guitar will only raise it up so much.

Defining the Boundaries of the 3D Stereo Field of Imaging

Consider this: the image of a sound never seems to appear further left than the left speaker or further right than the right speaker. Right? Right, unless the room is strange. (Sometimes unusual room acoustics can make sounds seem to come from odd places in the room.)

Remember . . . we're not talking about reality here. This is the world of "imaging." Because the exact placement is a figment of our imagination, different people see the left and right boundaries differently. Some say that it can't be further left or right than the speaker itself. Some people with quite active imaginations see sounds as far as a foot or two outside of the speakers. However, most people see sounds about 6 inches to the left or right of the speakers (depending on the size of the speakers). Check it out for yourself. Pan one sound all the way to the left or right and listen to see how much further the image seems to be beyond the speakers.

The left and right boundaries of imaging are shown like this:

Visual 14.
Left and Right
Boundaries of Imaging

When you turn the panpot (the left/right panning knob) you can "see" the sound moving left and right between the speakers.

Now consider front to back boundaries in regards to volume. Just how far do sounds recede into the background as you reduce the volume? How far behind the speakers is a sound before it disappears altogether? Try turning down one sound in a mix almost all the way so you can barely hear it, and see how far back behind the speakers it appears to you. Most people seem to imagine sounds only a few inches behind the speakers, depending on the size of the speakers.

There is a psychoacoustic phenomenon based on previous experience wherein certain sounds appear to be even further behind the speakers than the normal imagined limit. For example, if you place the sound of distant thunder between the speakers, it can seem to be miles behind the speakers. The sound of reverb in a large coliseum or a distant echo at the Grand Canyon might also seem to be way behind the speakers. This is a good illusion to remember when trying to create unusually expansive depth between the speakers.

As previously mentioned, when you turn a sound up, it appears to be more out front in a mix. But how far out front will it go? First, no matter how loudly you raise the volume of a sound, you can't make it come from behind you. In fact, sounds rarely seem to be more than a short distance in front of the speakers. Again, turn one sound in a mix up *really* loudly and see how far in front of your speakers you perceive the sound to be.

Most people imagine sounds to be only about 3 inches to a foot in front of the speakers. Again, it depends on the size of the speakers. A loud sound in a boom box will appear only about 2 inches in front, whereas sounds in huge PA concert speakers might appear as far out front as 6–10 feet.

Regardless of our perception of the exact limits of imaging from front to back, it is easy to imagine the placement of sounds from front to back, with volume being the main factor that moves a sound. Therefore, the normal stereo field is actually three-dimensional! I'll show the rear boundaries of imaging like this (the front boundaries aren't shown because they would just get in the way):

Visual 15.
Front and Back
Boundaries of
Imaging

Finally, what about the upper and lower limits of imaging? As discussed earlier, high frequencies seem to be higher between the speakers than low frequencies. The questions are: How high are high frequencies? And, how high do the very highest frequencies we hear seem to be between the speakers? Grounded people say sounds never seem any higher than the speakers themselves. The creative ones say sounds seem to float a few inches above the speakers. Pull up a high-frequency instrument like a cymbal or a synth sound that is extremely high like bells or chimes and notice how high it seems to appear between your speakers. Most people see really high-frequency instruments only a few inches above the speakers. Again, the exact limit depends on the size of the speakers and the imagination of the listener. Regardless of the exact limit, sounds never seem to come from the ceiling. Imaging is limited to somewhere around the top of the speakers.

Now what about the lower limit? Low frequencies commonly come through the floor to our feet. Therefore, the floor is the lower limit. The upper and lower limits can now be shown like this:

Visual 16.
Up and Down
Boundaries
of Imaging

No matter how far we pan a sound to the left, it will never sound like it is coming from much further left than the left speaker. Likewise on the right. We "see" sounds only a little bit in front of and behind the speakers. We don't hear sounds much higher than the speakers, but they do come through the floor.

The limits of imaging can be shown with this one visual:

Visual 17.
Only Place Mix Occurs

Speaker size affects the perception of the boundaries of imaging. With a boom box, we normally don't hear sounds more than a couple of inches left/right, front/back, or up/down beyond the speakers.

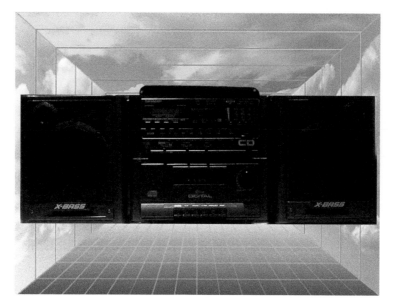

Visual 18.
Imaging Limits Around Boom Box

When listening to a huge PA at a large concert, the image might appear as far out front as 10 feet, and it might be 10 feet behind the speakers. It might easily be seen to be as much as 6 feet further left and right than the speakers themselves, and might be much higher and lower than the speakers.

Visual 19.
Imaging Limits Around Large PA

Even though our perception is a subjective experience and many people disagree . . . people do not disagree very much. Over the last 20 years we have asked students where they perceive the imaging around the speakers to be, and we have found that everyone's subjective experience is quite consistent. No one hears the sounds 10 feet to the sides of the speakers, or behind his or herself.

This is the space where sounds in a mix occur. In the world of imaging, sounds do not occur anywhere else in the room. Most importantly, you can see, *this space is limited.*

Therefore, if you have a 100-piece orchestra between the speakers, it's going to be crowded.

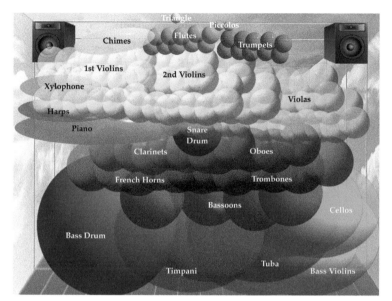

Visual 20.
Large Orchestra
Crowded Between
Speakers

You can't hear each individual violin in the mix because it is too crowded; you only hear a violin section. Whereas if you have only three violins, you can hear each one quite clearly.

Visual 21.
Three Violins with
Plenty of Space in
Between

Masking, where one sound hides or obscures another sound, is a major problem in mixing. If you have two sounds in the same place between the speakers, one of the sounds will often be hidden by the other sound. Because the space between the speakers is limited and masking is such a major problem in a mix, the whole issue of mixing becomes one of . . . crowd control!

As you can see, a sound can be moved around in the space between the speakers by changing the volume, panning, and pitch (equalization will make small up/down changes). These same three parameters are used not only to move sounds around between the speakers, but also to place and move effects, including delay, flanging, and reverb.

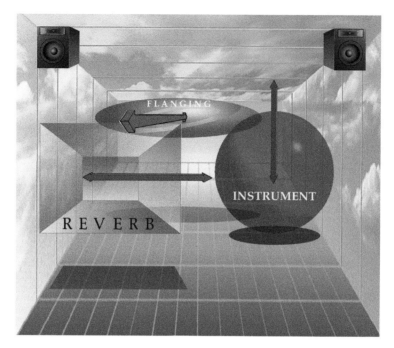

Visual 22.
Movement of Sounds with Volume, Panning, and EQ

A large part of mixing is simply to moving sounds to different places between the speakers in order to avoid masking so that you can hear each sound clearly. However, as I'll get into later, there is a bit more to it. Sometimes you just might want sounds to overlap, creating a fuller type of mix (instead of placing sounds apart from each other). You might want some sounds to overlap and others to be separate in order to highlight them. There is a huge number of possibilities.

As you might well be thinking, if masking is such a big issue, then it is important to know how much space a sound takes up in this limited world of imaging. In fact, not only do different sounds take up more or less space, equalization and effects can make a huge difference as to how much space a sound uses up.

Section C: Visual Representations of Sounds

Just how big is each sound in this world of imaging? With limited space between the speakers, you need to know the size of each member of the crowd so you can deal with the big problem of masking. The more space a sound takes up, the more it will hide other sounds in the mix.

Visual 23.
Solar Eclipse: Natural Masking

Limited Space

At first thought, you might think that you could make the space bigger by moving the speakers apart. The only problem is that the sounds become precisely proportionately larger in size, so you end up with no more space than you started with. On the other hand, a 3D sound processor allows you to place sounds outside of the space between the speakers. It does expand the space between the speakers. Surround sound (5.1 or any type of multichannel mixdown system) also expands the three-dimensional space between the speakers to include the entire room. These mixing techniques are explored in Chapter 8.

Size as a Function of Frequency Range

First, bass sounds seem to take up more space in the mix than high-frequency sounds. Place three bass guitars in a mix, and you'll have a muddy mix.

Visual 24.
Mud City

Bass sounds, being bigger, mask other sounds more. However, place ten bells in a mix, and you can still discern each and every bell distinctly from the others – even if they are all playing at the same time. High frequencies take up less space in the world of imaging.

Visual 25.
Ten Bells Playing at the Same Time

Therefore, the visuals representing high-frequency sounds are smaller and placed higher than the low-frequency instruments, which will be represented by larger images and placed lower between the speakers.

> Technically, it is very difficult to tell exactly where low frequencies, below 400 Hz, are coming from. Low frequencies are extremely difficult to localize between the speakers. Therefore, a more realistic visualization would have the low-frequency spheres less defined – they would spread out to cover the entire lower portion of the visual – creating even more masking. However, in order to be able to show the specific volume, panning, and EQ of bass, I will continue to use large, defined spheres.

Size as a Function of Volume

The louder a sound is in the mix, the more it will mask other sounds. Therefore, louder sounds are larger. A guitar that is extremely loud will tend to mask the other sounds a lot more than if it were soft. A bass guitar, already large, will hide other sounds even more when turned up loud.

Visual 26.
Loud Bass Guitar Masking Rest of Mix

Size as a Function of Stereo Spread

When you have a delay longer than 30 ms (1,000 ms = 1 second), you hear two sounds, which look like this:

Visual 27.
Delay Longer than
30 ms

An unusual effect happens when you put a delay on a sound less than around 30 ms. Because our ears are not quick enough to hear the difference between delay times this fast, we only hear one fatter sound instead of an echo. This effect is commonly called *fattening*. When you place the original signal in the left speaker and the short delay in the right speaker, the effect is such that it "stretches" the sound between the speakers. It doesn't put the sound in a room (like reverb), it just makes it "omnipresent" between the speakers.

A similar effect can be created by placing two microphones on one sound. Because sound is so slow (around 770 mph), you get about 1 ms of delay time per foot. Therefore, you will hear a short delay that will also create a stereo sound when the two mics are panned left and right between the speakers.

Visual 28.
Close to 1 ms Delay
Time Per Foot

Additionally, sounds in synthesizers are commonly spread in stereo with these same short delay times.

Visual 29.
Fattening: <30 ms
Delay Time

Just as you can use volume, panning, and EQ to place and move spheres, you also have control over the placement of the oblong sphere, or "line," of sound created by fattening. You can place the line anywhere from left to right by panning the original signal and the delayed signal to a variety of positions. The wider the stereo spread, the more space the sound takes up and the more masking it causes.

Visual 30.
Fattening Panned
11:00–1:00

Visual 31.
Fattening Panned
10:00–2:00

You can also bring this line of sound up front by turning the volume up . . .

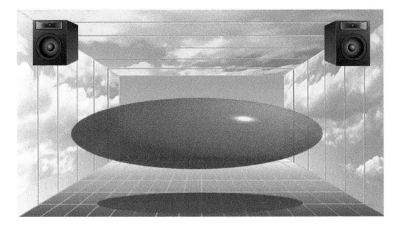

Visual 32.
Loud Fattening Right
Up Front

. . . or place it in the background by turning the volume down.

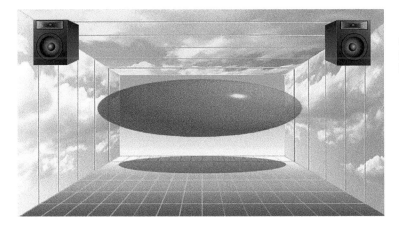

Visual 33.
Low Volume Fattening
in Background

You can also move it up or down a little bit with more treble or bass EQ.

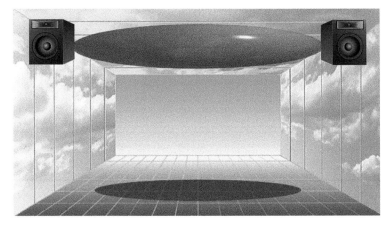

Visual 34.
Fattening with High-
Frequency EQ Boost

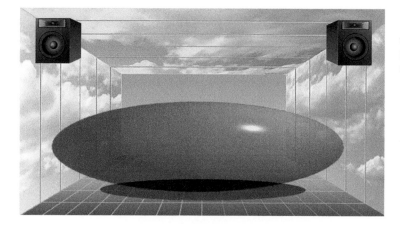

Visual 35.
Fattening with Low-Frequency EQ Boost

Size as a Function of Reverb

Placing reverb in a mix is like placing the sound of a room in the space between the speakers. A room, being three-dimensional, is shown as a 3D see-through cube between the speakers. Reverb is actually made up of hundreds of delays. Therefore, it occupies a huge amount of space when panned in stereo. It is like placing hundreds of copies of the sound at hundreds of different spots between the speakers. This is why reverb causes so much masking!

Visual 36.
Stereo Reverb on Sound

Just as spheres and lines of sounds can be placed and moved around in a mix, you also have control over the placement and movement of reverb with panning, volume, and EQ. You can place reverb anywhere from left to right by panning the two stereo outputs of the reverb in a variety of positions. The wider the stereo spread, the more space reverb takes up and the more masking it causes.

Visual 37.
Reverb Panned 11:00–1:00

Visual 38.
Reverb Panned 10:00–2:00

When you turn the volume level of the reverb up (normally done by turning up the auxiliary send on the sound going to the reverb), it comes out front in the mix.

Visual 39.
Loud Reverb

If you put reverb on a high-frequency sound, it will be a higher-frequency reverb. Same with reverb on a low-frequency sound – low-frequency reverb. Because low-frequency reverb takes up so much space, it is pretty rare to put reverb on low-frequency instrument unless they are by themselves in the mix. With EQ, you can raise or lower the placement of the reverb a little, which makes the reverb smaller (more trebly) or larger (bassier).

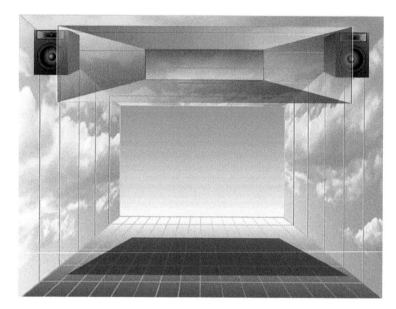

Visual 40.
Reverb with
High-Frequency
EQ Boost

Visual 41.
Reverb with
Low-Frequency
EQ Boost

These three basic sound images – spheres, lines, and rooms – can be placed within the three-dimensional stereo field between the speakers to create every structure of mix in the world.

Spheres represent sounds, oblong spheres represent fattening, and translucent cubes of light represent reverb. All other effects, including different delay times, flanging, chorusing, phasing, parameters of reverb, and other effects, are variations of these three images and will be described in detail in Chapter 3. With these various sound images, you can create a wide range of mix styles appropriate for various music and song styles. For example, you can create even versus uneven volumes . . .

Visual 42.
Even Volumes

Visual 43.
Uneven Volumes

. . . balanced versus unbalanced mixes . . .

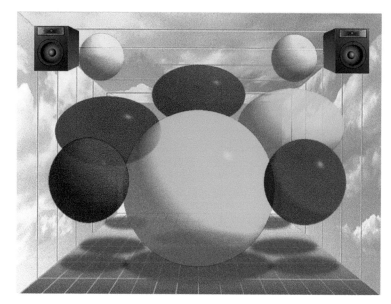

Visual 44.
Symmetrical (Balanced)
Mix

Visual 45.
Asymmetrical
(Lopsided) Mix

. . . natural versus interesting EQ . . .

Visual 46.
Natural EQ

Visual 47.
Interesting EQ

. . . and sparse versus full (wall of sound) mixes with effects.

Visual 48.
Sparse Mix

Visual 49.
Full (Wall of Sound)
Mix

This limited space between the speakers where imaging occurs is the stage, or pallet, on which you can create different structures of mixes. An engineer must be adept at coming up with any of the structures and patterns that can be conceived. Each one of these structures creates different feelings or dynamics. Just as a musician needs to explore and become thoroughly familiar with all the possibilities of his or her instrument, so must an engineer be aware of all possible dynamics that the equipment can create.

The art of mixing is the appropriate and creative placement and movement of these sound images. The mix can be made to fit the song so that the mix becomes transparent or invisible. Or the mix can be used to create musical dynamics of its own. It can be a tool to enhance and highlight, or it can create tension or chaos. A great engineer uses the mix to push the limits of what has already been done – without going off the deep end.

An engineer has the same range of control as the sculptor: both are working in 3D. In sculpture, the artist deals with shaping the images in a three-dimensional space. In photography and painting, the artist deals with composition and the way color tones relate to each other. In construction, the carpenter deals with first building a strong foundation. In Feng Shui, the consultant deals with the placement of elements in a 3D space. Here you are dealing with the Feng Shui of mixing.

In fact, what we are really dealing with the Sacred Geometry structures that affect us just like the archetypal geometrical shapes found in nature.

You now have a framework with symbols for each parameter of sound. Chapter 4 will go into the details of each piece of equipment in the studio. Chapters 5 and 6 use the visuals to discuss how each piece of equipment can be utilized in the mix to create all the dynamics that the "engineer as musician" wields. But first, I'll explore all of the reasons for creating one style of a mix or another in Chapter 3.

Notes on Design of Visuals

Shape

At first thought, a dot between the speakers might seem appropriate. When a sound, such as a vocal, is panned to the left speaker, the dot would move to the left speaker; the dot would move right to represent panning to the right. This is a common representation used by many people when discussing left/right placement of sounds in the stereo field.

A round image is most appropriate, especially when you consider the way two sounds seem to meet when they are panned from left and right to the center. When they are brought together and start to overlap in the middle, the image suggests that the sounds should be round and symmetrical. If you were to use an image of a guitar, the neck of the guitar would puncture the adjacent sound first because they are both panned toward the center, unlike the way two sounds actually meet and then overlap.

A solid dot has its faults, though. Sounds are not solid objects. Two sounds can be in the same place in a mix yet still be heard distinctly. Therefore, it makes sense to make the sounds transparent or translucent. If you use transparent spheres to represent the sound field of the image as it appears between the speakers, then two sounds can be seen and heard in the same spot.

Color

There are many systems for mapping colors to frequency. The most common is a mathematical calculation based on the frequency of the color in nanometers. Although many "intuitives" have also mapped colors to frequencies. There are even traditional colors for each chakra.

The primary function of color is to differentiate between different types of sounds. Different colors could be made to correspond to different sound colors, types of waveforms, or frequency ranges. But since I don't want to require people to learn such a system to be able to understand the visuals, I will only use color to help differentiate between sounds in the mix.

When harmonic structures and equalization are discussed, colors will be assigned to specific frequency ranges.

Visual 50.
Virtual Mixer EQ

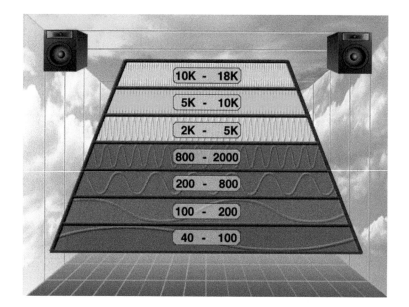

CHAPTER 3

Guides to a Great Mix

(Reasons for Creating One Style of Mix or Another)

So what makes a great mix? As a professional engineer, it is important to be able to answer this question. Of course, you know a great mix when you hear it. However, the key is to be able to define it so you know how to get it every time. It is also helpful to be able to articulate it so that you can explain to a potential client why they should hire you. But most importantly, it is important to know in your mind precisely what you think a good mix is so you can stand your ground when someone is telling you to do something that you know will make the mix suck.

Knowing what makes a great mix has its advantages. First, when you sit down at the console (or computer), it is absolutely critical that you conceptualize the mix before you begin. As Mom always said, "Before you begin, it is good to know where you are going." This is an important concept. *Don't touch a knob until you know where you are going.* Of course, sometimes, new ideas and directions might develop once you get into the mix. And it's also true that if you have enough time, sometimes you will get there through luck. However in a world where time is money, *you can't afford to not take the most direct and efficient route.* Conceptualize before acting.

Second, knowing what makes a great mix can help you communicate better with bands and groups. Many engineers – even great ones – know how to create great mixes, but they may not know how to explain them. They simply fly by the seat of their pants. As I'll discuss later, often conflicts arise in which you need to be able to explain why you believe something is right.

Sometimes it is just helpful to be able to explain things, because the people you are working with don't really know what they want. Even if they do know what they want, they often can't explain it because they don't know the terminology or what the equipment does. People normally appreciate it when you can help them navigate this complex world of equipment, music styles, and values. The more you help them to understand the complexities of a mix, the easier it will be to work with them.

Third, when a band is booking a session, being able to explain what makes a good mix can be pivotal to getting the job. Imagine that someone walks up to you and says they have $100,000 and would like to hire you to mix their project. But first, they want to know what you think is a good mix. They want to make sure that you are on the same wavelength. This can be tricky. If you like *clear* mixes, they might like *full* mixes. If you like a lot of effects, they might like no effects. No matter what you come up with, there is always the possibility that they will like the opposite.

However, in all this diversity, there is one thing that just about no one can argue. The mix should be *appropriate*: appropriate for the style of music, appropriate for the song and all of its details, and appropriate for the people involved – when the mix fits all three, the overall recording is much stronger and more powerful. Most bands will agree that the mix should be appropriate for the style of music. Just about everyone likes it when the mix fits the particular song and all of its details like a glove. And of course, everyone is happy when the mix fits their own tastes; and they are especially thrilled if the mix fits the tastes of a mass audience and becomes a hit.

> **The tricky part is**
> **balancing what the music, song details, and people**
> **are all telling you to do.**
> **The people are the wild card.**

These three aspects are valuable guides in choosing the type of emotional and musical dynamics that you create with the tools in the studio. Take a look at each of these guides in more detail.

Section A: The Style of Music

Over the years, each style of music has developed its own traditions as to how it is mixed.

This includes specific volumes, panning, EQ, and effects for instruments that occur in that style of music. For example, the volume of a kick drum in Big Band music is quite different than in hip hop or dance music.

When mixing by yourself on your own stuff, you can do whatever you want. Other than social disdain, there are no "mix police." However, when you are mixing for other people, they are commonly looking to have their mixes sound like the style of music they play.

The trick is to learn the traditions that exist for each style of music. Some traditions have become very strict. Classical, big band, and acoustic jazz have some of the strictest traditions. For example, if you don't mix classical music the way it has always been done, you could go to jail. It's against the law. Pull out a flange or turn up the kick drum too much on a big band mix, and it's over. On the other hand, hip hop, and especially dance music and electronica, have much looser rules. In fact, in these styles of music, often the rule is to break the rules. However, even these styles of music have developed their own traditions. The amount and type of bass EQ utilized has become quite strictly defined in most dance music.

> It's funny to think that if Beethoven and Mozart had decked out a recording studio full of equipment, they would have pushed it to the limits of its possibilities. You know those guys were kind of nuts – a little crazy. If they had samplers and Digital Audio Workstations they would have been doing some crazy stuff. They would have been turning sound waves around backwards, doing some bizarre things with loops, and they would have gone off the deep end with effects. However, since they didn't have these tools back then, we are now stuck with quite conservative rules for mixing classical music.

It is interesting to think about how different styles of music developed such traditions. How is it that a style of music came to be mixed a certain way? Why is big band music mixed differently from heavy metal? Who started these traditions anyway, and why did they decide the way they did? Music is an extension of a person's personality. A jazz player's music is often a reflection of their inner world. Likewise, a heavy metaler or hip hop artist's music is a reflection of their personality and the culture. Good mixes are normally just another extension of the music itself. Therefore, these mixing traditions are a direct result of the style of the person who plays the music in the first place.

Most bands do subscribe to having their music sound like a specific genre of music; however, there are aliens out there. Your next session just might be a big band that wants the Pink Floyd mix (with lots of effects and mixing tricks). It is critical to know just how strictly the band subscribes to having their mix sound like their style of music. You then know how creative you can get.

Then there are the bands who say they want a mix like no other. They want to create their own style of mix. This makes sense when it comes to music, but I've found that when it comes down to it, they still want the mix to fit their overall style of music.

Additionally, within each type of music, there are often numerous styles. Country music is a good example. In country music, there are at least twenty different typical styles of mixes, ranging from Hank Williams, Sr. to Taylor Swift. Furthermore, people who are into country music have often been listening to country all their lives, so they know when it doesn't sound exactly like what they are used to. Rock is the same way. People who are into a certain type of rock and roll know what their style of music should sound like, whether it be rockabilly, alternative rock, or heavy metal. But they can't necessarily tell you how to get the sound they want with the studio equipment.

The way that each style of music has been mixed throughout the history of recording often plays an important role in how a particular project should be treated. It is also helpful to listen and stay on top of current music industry trends. Each style of music goes through its own transitions over the years. Rap is a prime example. The types of mixes being done for rap music have changed dramatically over recent years.

Homework for the Rest of Your Life – Due (Do) Every Day

If you have a favorite style of music that you listen to all the time, start listening to other types on the radio. Listen for the differences in the mixes for each style of music.

Once you have finished the book, you might come back and reread this chapter. Some of the suggestions might make more sense to you once you are familiar with the details of what each piece of equipment in the studio can do.

Section B: The Song and All of Its Details

It is normally obvious to create a mix that fits the style of music. But the most important guide (often forgotten) is the song and all of its details. The details include the concept, intention, melody, rhythm, harmony, lyrics, arrangement, instrumentation, song structure, performance, hooks, and even the quality of the equipment. Although, you wouldn't think so, even the mix can affect the mix – for example, one aspect of the mix might make you adjust another aspect of the mix. Each one of the thirteen aspects could provide the reason for structuring the mix in some particular manner or creating a certain type of mix. Each aspect might prompt you to use one of the four tools (volume, panning, equalization, or effects) in a certain way. The mix might be used to enhance each and every detail found in the song or create tension with every detail. Regardless of how the mix interfaces with all of the components of the song, you should at least be aware of the relationship. It should be appropriate. Let's go through each aspect and see how it could affect the way a song is mixed.

Concept

The concept is a combination of the relationships of each of the other aspects, so it is one of the most important clues to the overall mix style. A multitude of various song concepts could lead us to create a wide range of different mix styles. For example, a song about chainsaw murders might be mixed with edgy EQ, some unusual cutting effects, and shocking dynamic volume and panning changes. Whereas a song about Universal love might be mixed with more natural EQ, balanced panning, even volumes, and nice, mushy effects, like reverb and long delay times with feedback or regeneration (where the delay continues to repeat).

Melody

The nature of the melody line can easily affect what an engineer does in the mix – overall and at any specific point in the song. For example, if the melody is a major component of the song, you might consider making it bigger and more prominent with time-based effects. You might turn it into a stereo sound with some type of fattening, or make it fuller with reverb. If the melody is simple (or boring, for that matter), it might be a good idea to spice it up with some flanging, chorusing, or phasing (depending on the style of music). Simple melody lines might be placed a touch lower in the mix. You will still able to follow it even if you can hardly hear it. On the other hand, if the melody is extremely busy, it might be better to use fewer effects, brighten it with EQ and turn it up more, all so you can hear the detail. Occasionally, engineers will change panning or effects based on what happens in the melody line. Perhaps a section of the melody might be highlighted by adding an effect or, even more effectively, taking an effect off. Likewise, a piece of the melody line might be panned to the center in order to accentuate it.

Rhythm

The nature of the rhythm has a direct effect on the mix. The busier the rhythm, the cleaner and clearer you make the mix so you can hear more of the detail. Instruments are normally equalized a bit brighter so that the detail of the rhythm is more distinct. The volume of a complex rhythm part might be boosted just a bit in order to make the details clearer. You don't want to cloud the precision of an intricate rhythm, so use fewer time-based effects because there is not enough room for the additional delays.

On the other hand, if a rhythm is more basic – slow and simple – there is more room for time-based effects, like delay, flanging, chorus, and reverb. You can also add more bass EQ and turn the volume down a bit.

Harmony

The nature of the harmony parts and how they fit into the overall song also contributes to a different handling of the mix.

Differences in the number of harmony parts and their chord structure can provide important clues as to what might be done in the mix. For example, the more harmony parts, the wider the sounds might be panned. A single harmony part is rarely panned all the way to one side. The type of musical chords that are created with the harmony parts could affect their placement. A dissonant type of chord might be placed back in the mix; a sweet, angelic chord might have the low-end rolled off with EQ and be mixed with spacey delays and reverb. If you want harmonies to be more cohesive, you might add fattening (short delay less than 30 ms). If you want them to be extremely homo-geneous, pan them all to one side and add the delay to the other side.

When the harmonies are not harmonies but simply the melody sung in unison, they might be spread wider in stereo to make them sound fuller and bigger. And the volume might be lowered because they might not be interesting enough to be put right out front. Of course, the quality of the performance will affect how loud you place them, but that will be taken up in the "Performance" section. You will not go to jail if you don't follow any of these suggestions. None of these ideas are rules; they are simply guidelines for creating a mix that better fits the song details.

Lyrics

Lyrical content is a major guide to how a recording engineer mixes a song. The nature of the lyrics often affects the overall mix because they help set the tone of the song. Particular lyrics can often trigger the engineer to create and place various effects in the mix. A line such as "living on the edge" is just begging to be panned far left or right in a mix. A line about psychotic behavior might make you boost some irritating frequencies on a guitar or keyboard. A line like "in the halls of love" might call for some sort of hall reverb effect. Effects might also be removed based on the lyrics. For example, it is common to take off effects (especially reverb) when you want to highlight an important line. It makes them more personal, up front, and close to your heart.

You should always listen closely to hear what the lyrics might be telling you do with the mix.

Density of Arrangement

The density of the arrangement is often a valuable clue as to how to mix the song. If an arrangement is extremely full, then there are two different plans of attack: weed it out or fill it out even more.

If you are trying to weed it out, you obviously don't want to make the mix fuller by adding additional effects. The idea is to try and make the mix appear cleaner and clearer so that you can hear the myriad details within the arrangement. Besides using fewer effects, sounds are commonly EQ'd brighter overall. When there is a large number of sounds and notes in the mix, the higher frequencies are masked more. As low frequencies take up more space between the speakers, they might be turned down a bit on each sound (especially in the muddy range from 100 to 300 Hz). Brighter EQs and less mudd make the sounds take up less space, so there is more room for each sound to be heard. You might also pan things wider with a busy arrangement. With so much going on in this limited space between the speakers, it often becomes necessary to utilize the entire stereo field.

On the other hand, occasionally, it might be appropriate to use the mix to help fill out a dense arrangement even more, making the massive "wall of sound" effect even more pronounced. By adding time-based effects (like delay, flanging, or reverb), you are actually adding more sounds to the mix. Perhaps a little extra low end might now be appropriate. You might not have a clear mix, but it will be so HUGE AND MASSIVE that no one will notice.

If an arrangement is sparse, there are also two approaches. You could help keep the arrangement sparse, which might be perfectly appropriate for certain styles of music, like traditional jazz or folk music. Or you can use the mix to fill out the space between the speakers. As mentioned above, you can use more time-based effects and fuller low-end EQ to fill out the mix. When there are fewer sounds, you have more room to play around with various effects.

Instrumentation

If the quality of an instrument sound is incredible, use the mix to highlight it. *Incredible* might mean really interesting, unique, exceptionally beautiful, very intense, or powerful. The complexity of the sound might also be its value. You can highlight the sound by turning it up and brightening the EQ, or you might make it into stereo with a time-based effect. If it's really cool, show it off.

On the other hand, unique and interesting sounds can be quite intriguing when low in the mix so they just seep through. Sometimes it might be appropriate to make a sound stereo with a short delay, and then place it in the background.

If the sound is not happening, don't highlight it by turning it up too much. You might not brighten up the EQ quite so much. Instead, you might put an interesting effect on it so the effect itself becomes the highlight. If you can't polish it, bury it a bit (but not too deep – then the mix will be wrong).

Song Structure

The structure of the song often affects what an engineer does in each section of the mix. Some bands will actually create a structure where each section of the song is completely different from the previous section. Frank Zappa, Pink Floyd, Mr. Bungle, and even The Beatles had songs where the mix was drastically different from one section to the next. It is a good idea to be on the lookout for bands who have created songs in which you could create different mixes to accent each section of the song – just because these types of mixes are so much fun to do.

Even when different sections of a song are not that different, you might subtly accent each section a bit differently. For example, a chorus will sometimes have a bit more reverb on the vocals. A lead break is often spiffed up by boosting the volume of the kick, snare, or bass and sometimes by adding more reverb to the snare. The same is commonly done in the vamp at the end of the song when the band is rocking out (or doing whatever they are doing). A bridge section might have a different type of reverb or a different type of panning on the instruments in order to create some variety. Creating subtle variations in the mix in different sections of the song creates more depth, so the closer the you listen to the song, the more you hear. You've all heard songs like this before. It makes it so you can listen to them over and over. Songs with depth!

Performance

The performance is often one of the most important aspects affecting the mix. The better the quality, the louder and more present the EQ (within reason). A great performance might require the sound to be spread left and right with fattening to really show off the talent. On the other hand, if it isn't incredible, don't put a spotlight on it by turning it up with no effects on it. At least turn it down a little in the mix or put a bit of reverb or some other time-based effect on it to smooth out (or hide) the rough edges. Again, don't bury it too deep. Don't put too much reverb on it, and don't turn it down too much. Then, not only do you have a bad performance, you have a bad mix. If the perform-ance gives you no goosebumps or chills, you just might use an effect to provide the goosebumps or chills.

Besides quality, a specific aspect of the performance could lead to specific mix settings. If you have a performance with timing issues, you might fix them with editing. Certain programs, like Pro Tools, have the capability to find the beat and even quantize the digital audio. If there are pitch problems, a chorus or flange effect will help cover it up. You can use a pitch corrector to fix the pitch. You can use little boosts in volume to simulate emotional intensity in the performance. Likewise, you can ride the volume faders up and down to smooth out a performance that fluctuates too much dynamically. If the performance has technique problems, you can sometimes use processing or effects to hide them.

Occasionally, a particular lick or riff will induce the engineer to pan it from left to right. You might also add more effects as a performance gets more intense. Or, on the other hand, you might gradually take off effects to match the performance as it gets more up front and personal.

Quality of the Recording Equipment and the Recording

If you have cheap equipment, you shouldn't make the mix too bright and crispy because it will show off any noise and distortion. With better equipment, you can often make your mixes cleaner and clearer. If you get good levels on your multitrack, and you set the gain structure of your effects correctly, you won't have to deal with it in the mix. Also, if you take the time to find the right spot in the room for the instrument and the right mic, you won't have to fix it in the mix later.

Mix

As mentioned, you don't ordinarily use the mix to tell you how to mix, because, of course, there is no mix in the beginning. However, often, doing one thing in the mix, might require that you do another thing. For example, when you turn a sound down and place it in the background, it is often brightened up more with high EQ because sounds in the background are less distinct. If you add a short delay or reverb, often the sound must be made a little brighter because the effect obscures the sound. Also, the original sound with the effect on it must often be turned down simply because two sounds are louder than one. When you put a delay on a sound, you might then pan the sound and delay separately. Occasionally, you might need to brighten a sound that has been compressed. Therefore, the mix can be affected by the mix.

Summary

Often, one aspect will dictate the mix more than the others. Commonly, it is the lyrics and the performance that play the biggest parts in determining what is done in a mix. However, this can vary drastically from song to song.

The key here is to simply listen to what each one of the thirteen aspects in the song is telling you to do in the mix – the more you make the mix fit the details of the song, the more powerful the song will be. This way the essence of the song is more likely to come through.

I used to work with a producer named Ken Kraft. As his engineer, I would often come up with unique and creative ideas. Just about every time, Ken would say, "What's the point?" After a while, I realized that he was really asking, "Does it fit the song? Does it get in the way of what the song is trying to say or does it enhance it?" Sometimes, I would realize that it didn't fit and say, "I know what you mean." Other times, though, I would explain to him how it fits with the song to make the song stronger. Occasionally, I wouldn't know how to explain it, so I would say something like, "I don't know why, but I *really* feel that this effect will work well with the song." And he would say, "Cool, let's try it out."

> When changing volume faders, panning, EQ, or effects settings
> ask yourself, "Does it fit?"
> If you can remember to do this at every step of the mixing process
> then you will always end up with a mix
> that is appropriate, the most powerful, and the best mix possible. The truth is that every single detail of the song is telling you how to mix the song.
> LISTEN TO WHAT THE SONG IS TELLING YOU.

Homework for the Rest of Your Life – Due (Do) Every Day

Every time you listen to a song, see which one of the thirteen aspects is playing the most important part in the creation of the mix. Over time, you will start detecting patterns for different styles of music and different songs.

Section C: The People Involved

Who's Mixing This Anyway?

One of the trickiest and most important jobs of an engineer is to balance your own values with the rest of the people involved. This includes the band, the producer, and even the mass audience. It often requires some serious diplomacy.

I'm here to let you know that it is important that you, as the engineer, are in control. Don't get me wrong! This does not mean that you dominate and don't listen to others' ideas. The professional engineer listens to all ideas and compares them with his own values and the values of the mass audience, then he makes a decision as to what is best for the song.

When I first started mixing, I wanted to please the clients, so I would ask them every step of the way if they liked what I was doing: "Anything you would like to change? Do you like this EQ? Do you like this balance of volumes? Do you like this panning? Do you like these effects? Do you like this overall mix? Are you sure?" I figured that if I went with whatever the band wanted, then they couldn't blame me for a bad mix. However, my mixes kept coming out mediocre. The problem was, I was letting the band mix it. The worst case is when the band comes back a couple of weeks later unhappy or displeased with the mix, having forgotten they approved everything each step of the way.

> Once you have finished reading this book
> and have gotten some experience,
> you will know how to mix
> better than most of the people in the room.

Most band members don't know all that can be done in a mix. Of course, they know what they like when they hear it; but they don't know all the parameters of all the equipment, so they don't know how to get it.

Professional engineers don't let the band perform the mix. Doctors don't let patients do the surgery. But doctors damn well better listen to the patients' concerns. The same goes for engineers.

If you are going to be in control, there is a responsibility that goes along with it. *You must give total respect and consideration to every single idea that someone comes up with.* This doesn't mean that you have to use the idea. But if you don't give it full consideration, two things might happen. First, you might upset someone. Second, they might shut down and no longer share any creative ideas. When you give an idea total respect (even if you put the kibosh on it later), it leaves the door open for more creative ideas to flow. Keep the door open.

Obviously, you are probably thinking, How do you tell someone their idea sucks, but with respect! The key is to use reasoning that makes complete sense. This means you must develop a repertoire of *real reasons* as to why you say something should be the way it is in the mix. Your reasons must be realistic and logical.

Telling a band that you have been mixing music for 20 years (even if you have) doesn't work. Telling them "Trust me" doesn't work. Telling them you have read and studied *The Art of Mixing* doesn't work. Even telling them you've mixed twenty platinum albums (even if you have) doesn't work. You must have specific reasons that make sense.

The classic problem arises when you have been working on a mix and say you've got it sounding really good. In fact, you might even be seeing visions of Grammies in your head. Then you hear someone in the back of the room say, "Excuse me. Could you please make the mix suck by turning up my instrument too much?" You turn around and look at them (as visions of Grammies fade into the distance) as they say, "I'm paying for this." Your heart sinks as you think, "So much for the Grammy."

It is so easy to give in and say, "OK." You might think, "It's not my fault if the mix sucks. At least I'm getting paid." However, the problem is that your reputation is at stake. It is your name that is going on the album – and you can't put a disclaimer on the cover. But most importantly, you just want it to be as good as possible.

The worst case is that you give in, and driving home you realize that you were right. You can't even listen to the mix because it does suck.

> The trick is to figure out
> the *real reason* as to why you were right
> and how you could have explained it to them
> so that it made sense.

Once you figure it out, don't forget it! The worst thing that could happen is a few months later, a totally different band comes in and again someone in the room says, "Could you please make the mix suck, exactly like that last band?" *And, you still don't know how to explain to them why they are wrong . . . .* That's the very worst that could happen.

> The job of the professional engineer is
> to develop a repertoire
> of reasons for doing what you do
> that makes sense
> so that you can
> help the band to understand
> why the mix should be a certain way.

The ideal engineer is one who not only knows what is best for the project, but also knows how to explain why to the clients.

Instead of dealing with this predicament, some engineers simply don't allow the band in the room during the mix. The only problem is that you are losing out on any creative ideas that the band might have. And, having been working with the song forever, the band often has many very cool ideas that are perfectly appropriate for the song and totally cool.

It is important to point out the difference between "preference" and "wrong."

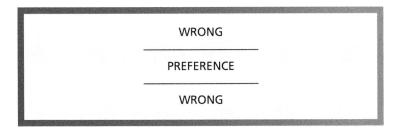

> WRONG
> ——————————————
> PREFERENCE
> ——————————————
> WRONG

There is a large world of preference, where one idea may be a bit better than another idea in one person's mind or another's – however, they are both very good. It is when the ideas move into the realm of "wrong" that you need to go into action and pull out those valid reasons. "Wrong" is defined as ideas or mix settings that have become socially unacceptable. Even babies and grandmas object. In this range, any professional engineer would agree that it is not acceptable, and just about everyone on the streets would also agree.

In the world of preference, it is still good to make your argument. You might decide, more or less, to put up a fight if you feel strongly enough about it (choose your battles carefully). But you keep in mind that you are in the realm of *preference*, so if you lose, it isn't that big of deal. In the realm of preference, there is no need to be so attached to the outcome. Besides, it takes all types for the world to go around.

However, when you are in the realm of "wrong," you need to put up a good fight – with total respect and armed with reasons that make sense.

When you know you are right, you should try to make a case for it; but be sure you are right. The absolute worst thing that could happen is to convince the band that you are right when you aren't. The band is going to listen to the project hundreds of times after they leave the studio – in a wide range of places, with all different kinds of ears. Later, they will know for sure if something is not right. Therefore, if you are going to try and convince someone that you are right, you had better be right.

The experienced engineer must attain a highly developed set of values to justify making certain decisions. Once you have gained the experience to really know what is right, you can then command that respect. When an engineer *really* knows, it often comes across in his or her demeanor. At this point, the band will often defer to the engineer. Some engineers have paid their dues and are so experienced that they have earned the right to be respected. The problem occurs when a recording engineer's ego or intensity is not backed up by experience.

This book is one of your key tools for learning the specific reasons why a mix should be a certain way. In a session, I will commonly refer to the basic concept that there is a limited space between the speakers, and if you make any one sound too big and overwhelming with volume, EQ, panning, or effects, it will make the other sounds smaller by comparison. It is all relative, and there is only so much space. Here are some other examples of specific reasons.

Say the band is asking you to turn up the kick drum too much. There is a wide range of levels that the kick could be at and still remain within the realm of preference. The problem comes when it is obviously outside the realm of preference. You might say something like, "You know, for this style of music and this type of song, it is rare that a kick drum is ever this loud in the mix. Do you really want it to be that loud?"

Or say the singer wants you to turn up the vocals volume. Again, you might point out that for this style of music, it is almost abnormal to have the vocals that loud. You might also point out that when the vocals are turned up that loud, it dwarfs the rest of the band, making it sound wimpy. Ask them if that's what they really want.

When doing rap or hip hop, the band often wants the 808 rap boom sound to be so loud that it rumbles the windows. This is usually because they are used to listening with the bass EQ cranked all the way up (either in their car, at home, or at clubs). Therefore, in the studio, the rap boom might not seem big enough to them without this bass boost. Just pointing out how it is going to be boosted more with EQ in the real world can make them realize that they shouldn't boost it too much in the studio – otherwise it could seriously blow up speakers later.

> In case you aren't familiar with the term *808 rap boom*, it is the extremely low-frequency sound that is used as a kick drum. You've probably heard it many times in the car next to you! It is called the 808 because it was originally used in the Roland 808 drum machine. Now, everyone and their brothers have sampled it, and it is available in a wide range of drum machines and synthesizers. The sound is actually a recording of a marching band bass drum. Also, if the sound changes pitch, it is then considered a bass instead of a kick!

Another example involves the reverb level placement in a mix. The problem occurs when you listen to reverb in solo. Once you have heard it alone, your mind recognizes it better when it's in the mix. Therefore, it seems bigger in the mix than it did previously. Because of this psychoacoustic effect, the band often wants the reverb lower than what you commonly hear on the radio. Explaining this phenomenon to them can help you get reverb up to a more appropriate level and help them understand the reason why.

Here's one final example. Perhaps the band wants to put an effect, like reverb or delay, on a sound like a bass guitar or kick. You might explain that bass instruments already take up a lot of space in a mix. Because reverb is made up of hundreds of delays, it really takes up a lot of space. When reverb is extra bassy, it takes up even more space, thereby masking other sounds tremendously. Explaining this to the band will help them realize how much these sounds are masking the other sounds in the mix.

As you can see, it becomes critical to truly understand the dynamics that go on in a mix so that you can logically explain to a band why one move makes more sense than another. However, if someone is still adamant about their ideas after you have discussed everything, you can try both ideas (yours and theirs), and maybe they will hear your point. It is important to remember that the band may not be able to explain why they want what they want, but their ideas are still cool. You just might hear their point and be able to make a compromise adjustment that satisfies everyone.

As a last resort, you can suggest making two mixes. Only do this if it is absolutely necessary because it takes more time.

Values of the Engineer

It could be argued that it is better for an engineer to use personal experience and intuition to set new trends instead of following them. And if you can feel it, go for it.

As discussed, the people who first started doing mixing didn't have any traditions to base their mixes on. They were just going on inspiration and what felt good to them. They had no reference point. They had to trust their feelings every step of the way. Perhaps there was a small component of tradition in that they would try to make it sound in the control room like it did in out in the studio. However, for the most part they were basing it on what felt right. These days, for the most part, we come up with creative ideas by taking other's ideas and putting them together to create something that is unique. In the beginning, they were producing the ideas out of thinner air.

Now there are thousands of ideas out there, and hundreds of little techniques and tricks that recording engineers have come up with. And just like the rapper who is sampling others' material and creating something totally new, we now can come up with some incredible mixes by simply utilizing the ideas of others. You could say that a great engineer is often a master thief. The good news is, there is absolutely nothing wrong with stealing another's mixing ideas. You definitely won't go to jail. In fact, the more you steal, the wider the repertoire of ideas you have at your fingertips to utilize on the next mix you do. Therefore, steal, steal, steal.

> As some say, "It's all been done before," and we are only rearranging all that has been done before. On a subconscious level, you could say that everything has already been created. We are only remembering it. There are only so many notes that can be put together in so many ways in music. There are even fewer ways that the four tools of volume, panning, EQ, and effects can be put together to create every mix in the world. Therefore, go ahead and access the well and use what you need. True creativity comes when you start to use combinations of these magic tricks (that you have stolen, or not), to create something totally new.

> As an engineer,
> you need to develop a bag of tricks.
> The source doesn't matter.

> I would say to create music or a song based on a formula (such as the "boy" bands) is evil. Some would disagree. It is true that producing formulaic music is nothing compared to murder and war. However, to steal mixing ideas is not evil at all. (You're not really stealing ideas, you are simply borrowing them).

Professional engineers have a bag of tricks that is overflowing. There are different categories of tricks. First, there are the tricks you use over and over, regardless of the style of music or song. For example, I use fattening (delay less than 30 ms) and a longer predelay on reverb quite often. The second of type of tricks are the ones that you only pull out of your bag occasionally and when it's appropriate. Preverb (reverb before a sound) or extremely long predelays on reverb are examples.

Then there are the types of tricks that you invent yourself (or at least you think you have invented . . . you may have just never heard anyone else do it). These are the tricks that you keep dear to your heart, of course. If you develop enough of your very own tricks (and they aren't too "out there"), then you might be on the way to creating a new tradition. Even if you have stolen ideas, over time, they will become your very own. The key is to develop your own style based on your very own bag of tricks.

Begin developing your own bag of tricks by finding one or two reverb presets in the unit you are using that are your favorites. These would be reverb settings that work quite well over and over. Find one for drums and one for vocals and other instruments, like guitar or keyboards. Then, as soon as you have the bandwidth (in your head), add more presets to your memory (in your head). In this way, you can begin developing your own bag of secret weapons. Then always be on the lookout for any ideas that you can absorb from radio and CDs. After a few years your repertoire of little tricks should become quite large.

Values of the Clients

The experienced engineer knows the importance of paying special attention to other people's ideas, even if they are out of the ordinary (or completely nutty). It is important for the client to realize that you care about their ideas. The trick is to get very quick and sharp at weeding out bad ideas (or less than great ideas) from good ones without hurting their feelings.

However, the band and the songwriter do have a major advantage over you as the engineer. First, they have already spent a lot more time with the song and know it much more intimately. Fresh ears on a song are nice, but intimacy helps. Second, the songwriter might have ideas that no one else could possibly come up with because he or she is so intimately involved with the creation of the song. If we think of a song as an extension of a person's personality – of his or her feelings and emotions – then it makes sense that the person who wrote the song would have more cohesive or holistic ideas for the mix.

It is the job of an engineer to pick up on the heart and soul of the song, the feel, in order to create a mix that is most appropriate for the song – whether that means sweetening it or creating tension. Paying close attention to the band and the songwriter's ideas can help you access this heart and soul.

It is, therefore, important to figure out the values and desires of the client. Engineers often ask the client about their values and then listen closely for any clues as to what kind of mix they might like. Assuming they are within the realm of preference, discuss your ideas with the band. Again, listen to their ideas, but remember you are the engineer. If your ideas differ, tell them why with specific valid reasons.

It is important to always judge every single idea that comes out of anyone's mouth based on its own merits. The problem is when someone who is extremely inexperienced or unclear about the recording and mixing process, out of the blue, comes up with a completely ingenious idea. Often, bizarre requests, seemingly devoid of any reason, can be pure genius. In fact, I imagine that whoever first worked with David Byrne must have wondered about him at first. To quote a few lines, "Don't judge a book by its cover," and "Out of the mouths of babes can come true wisdom." In other words, never judge someone based on preconceived ideas of who they are. Meet them on a creative level. Genius can easily be masked by nervousness.

On the other hand, you can't always count on someone who is extremely experienced and even has an incredible ear to always come up with ingenious ideas. You never know when the next thing that comes out of their mouth will be nonsense. Nonsensical ideas are often based on ego or insecurity, but sometimes it is just simple inexperience.

The whole key here is to always take every comment or suggestion at face value. Regardless of the source, ask yourself:

> "Does it fit?
>
> Does it distract from or enhance the essence of the song?"

And if it doesn't quite fit, is there some type of derivation that could make it fit so it accomplishes the creative goal behind the idea? Remain on your toes at all times – ready to weed out the genius from the B.S. with one fell swoop of logic and aesthetic values. Actually, it is often more like a slow tug of war than a fell swoop.

The truth of the matter is that if you simply gather all of the ideas from everyone involved in the project, you will end up with a plethora of cool ideas. In fact, as the engineer, you should be gathering these ideas from the second the band walks in the door. Whenever anybody (including yourself) comes up with a good idea, store it in your creative bank. Write them down, so you don't forget any of them.

You should be on the lookout for any good ideas that pop up throughout the session no matter how small or off the cuff. You might overhear someone talking to someone else saying that they would like to put an echo at the end of one of the vocal lines. Snag it out of the air and put it in your creative bank. What commonly happens is that during mixdown, both you and the band will forget the idea. Then, a couple of weeks after you've mixed the song, the band member who had the idea will be listening to the project and say, "Dang it, the engineer forgot to put that effect on the vocal. I'm going to another studio next time." No matter how off the cuff, don't forget any idea. Of course, you don't have to use every one, but it's nice to have a bunch of them to choose from. Don't forget to gather your own ideas as well. Put them in your creative bank, so you also can cash them in on the mix.

Values of the Mass Audience

Often a band comes into the studio, and they just want their music to sound like a hit. Really, what they are asking is, Does it fit the values of the mass audience? Some people see this as blasphemy, selling out, and the death of pure heart and feeling in music. This may be true for songwriting, but in mixing, this is not necessarily the case. What the band is saying is they want it to fit the high-level values of the mass audience that we are used to hearing on the radio and CDs. Often the most unique and creative types of mixes appeal to the largest audience. Many hits have incredibly cool and creative mixes. Some of us would be very happy to be able to create mixes similar to what is currently on the radio and CDs.

As an engineer, it is important to stay on top of trends in mixing. The world of dance music (electronica, techno, trance, house, etc.) is where most of the new trails are being blazed. You can often use ideas from other styles of music in your own favorite style of music.

Tradition Versus Inspiration and Creativity

As discussed, if inspired, do whatever you want. Who knows? You might be the next big hit. However, normally a good engineer is one who balances tradition with creativity. Once you learn the exact limits of the traditions for each style of music, the trick to being creative is to push the limits. However, if you push them too far, then you might easily be thought of as an inexperienced engineer, just plain nuts. Here are the limits of tradition:

```
┌─────────────────────────────────────┐
│                                      │
│              ───────────             │
│              TRADITION               │
│              ───────────             │
│                                      │
└─────────────────────────────────────┘
```

You can only push the limits so far until you are considered "off the deep end."

```
┌─────────────────────────────────────┐
│                                      │
│                 NUTS                 │
│            ──────────────            │
│              ACCEPTABLE              │
│            ──────────────            │
│               TRADITION              │
│            ──────────────            │
│              ACCEPTABLE              │
│            ──────────────            │
│                 NUTS                 │
│                                      │
└─────────────────────────────────────┘
```

By just pushing the limits of tradition enough, you help to change the world.

When I first started mixing, I wanted to change the world. I was extremely creative and was often coming up with mixes that were off the deep end. After a few years of bands coming back and asking for a more "normal" mix, I finally reeled myself in. I realized that if I was going to make it in this industry, I would have to be a little more conservative. In order to maximize my fun and creativity, I began paying close attention to the band and their music to see when going off the deep end might be appropriate. I'm still always looking for such bands to walk in the door.

Great mixes are normally a balance between tradition and inspiration or creativity.

> The "venue" where the music will eventually be played on is another minor but important factor that will influence the way it is mixed. Most projects will be played on a wide range of systems. If the project is going to play on a cheap car radio, it is important that there is enough bass in the low mids. If it is played on a killer system, it should accommodate. Mixing for TV is different than mixing for movies. When mixing for movies, you might add more sub-bass at certain critical moments. When mixing dance music for a rave (or party), you might bump up the bass. You are locked to the container that the project will play on, so take it into consideration.

It is obvious that the mix should fit the style of music – in fact, it almost goes without saying. But in addition to the style of music, the more the people involved pay attention to the details of the song as the primary guide in determining the mix, the more cohesive the mix will be. The mix is normally much better when everyone involved is basing their opinions on the song instead of their own personal desires. It's really great when everyone in the room is listening to what the song is telling them to do in the mix. Yet you never know when a person might have some inspiration from another world – whether it comes from God, angels, or aliens – that is pure genius. Such ideas might be more appropriate than basing the mix on the song itself. But, only if they are appropriate.

When all is said and done, the truth is that the precise best mix that you can end up with is all laid out before you from the beginning. The style of music tradition dictates certain volume, panning, EQ, and effect settings. Then, each of the details of the song dictate more refined volume, panning, EQ, and effect settings (and movement). The people involved normally just screw it up, unless they are also in tune with what the song is telling them to do in the mix. However, sometimes people involved get profoundly creative ideas from who knows where.

CHAPTER 4

Functions of Studio Equipment and Visuals of All Parameters

To simplify all of the functions of a huge variety of studio equipment, I have broken down the equipment into categories based on the function of each piece in the recording studio:

1. Sound creators: all instruments, acoustic to electric, and voice to synths.

2. Sound routers: mixing boards, patchbays, and splitters.

3. Storage devices: hard drives, tape players, sequencers, and samplers.

4. Sound transducers: mics, pickups, headphones, and speakers.

5. Sound manipulators: processing and effects.

Sound creators range from acoustic to electric instruments, from voice to synthesizers.

Visual 51.
Sound Creators

Sound routers route sound from one place to another. Mixing boards route the signal to four places: the multitrack, the monitor speakers, cue headphones (for the band out in the studio), and the effects (so we can have a good time). Patchbays are just the back of everything in the studio – the back of the mic panels, the back of the multitrack (inputs/outputs), the back of the console (ins/outs), and the back of the effects (ins/outs) – located next to each other so we can use short cables to connect them.

Visual 52.
Sound Routers

Storage devices store sound or MIDI information and play it back. Hard drives store digital sound, and MIDI information. Tape players store digital or analog sound.

Visual 53.
Storage Devices

Sound transducers take one form of energy and change it into another. Microphones take mechanical energy, or sound waves, and change it into electrical energy. Speakers take electrical energy and change it into mechanical energy, or sound waves. Likewise, pickups on guitars take the movement of the strings and change it to electrical energy.

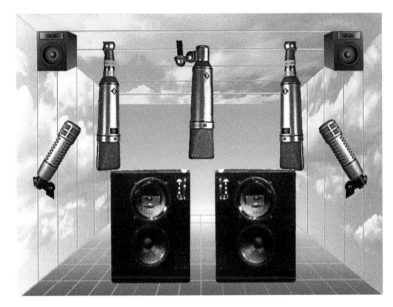

Visual 54.
Sound Transducers

Most of this chapter will be spent on sound manipulators. This includes processing that is used to control a sound, or effects that add an additional sound to an existing sound.

There are three components to sound: volume (or amplitude), frequency, and time. That's it. Therefore, every sound manipulator used in the studio can be categorized as to whether it controls volume, frequency, or time – or a combination of these. Note that the following chart categorizes each piece of equipment based on the *primary* function that it is designed for. For example, reverb has a volume control on it, but you don't buy a reverb unit to change the volume of a sound. You buy it to create the time delays that we know as reverb. Therefore, it is under "time" only. Likewise, a compressor/limiter has time functions on it, but you buy it to control the volume of a sound.

Visual 55.
Effects Rack

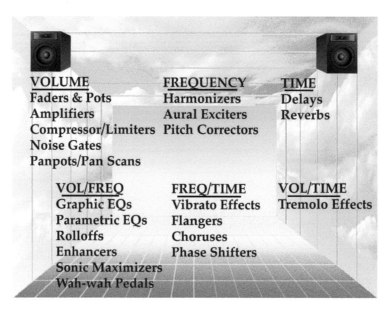

Chart 1. All Sound Manipulators

Section A: Volume Controls

Faders

Volume faders control the volume of each sound in the mix, including effects. The level set for each sound is based on its relationship to the rest of the tracks in the mix. When volume is mapped out as a function of front to back, you can place any sound or effect up front, in the background, or anywhere in between by using the faders.

However, the level that you set a sound in the mix is not based solely on the fader. If the level of the faders was the only thing that affected the volume of a sound in a mix, you could mix without even listening. You could simply look at where the faders are set on the console. There is more to it than that.

When you set volume relationships in a mix, you use apparent volumes to decide on the relative balance – not just the voltage of the signal going through the fader. The apparent volume of a sound in a mix is based on two main things, fader level and waveform, and another minor one, the Fletcher/Munson Curve (see the "Fletcher/Munson Curve" section below).

Fader Level

First, the level of the fader does affect the volume of the sound. Change the level of the fader, and the sound gets louder or softer. When you raise a fader on a mixing board, you are raising the voltage of the signal being sent to the amp, which sends more power to the speakers, which increases the sound pressure level (SPL) in the air that your ears hear. Therefore, when you turn up a fader, of course, the sound does get louder.

The decibel (dB) is used to measure the amplitude of the signal at each stage of this circuit. In fact, there are very specific relationships between voltage, wattage, and SPL. Decibels are the main variable affect the apparent volume of a sound. However, there is another important factor: the waveform of the sound.

Waveform (or Harmonic Structure)

The *waveform*, or harmonic structure, of a sound can make a big difference as to how loud we perceive a sound to be. For example, a chainsaw will sound louder than a flute, even if they are at exactly the same level on the VU meters. This is because the chainsaw has harmonics in the sound that create a dissonant chord, which is irritating – or exciting, depending on your perspective. These odd harmonics are edgy to our psyche, which make them seem louder to us. Therefore, a screaming electric guitar will sound louder than a clean guitar sound, even if they are at the exact same volume in the mix.

The Fletcher/Munson Curve

A minor factor contributing to the apparent volume of a sound is the Fletcher/Munson Curve. The biggest problem with the human hearing process is that we don't hear all frequencies at the same volume – especially those at low volumes. Fletcher and Munson did a study that shows just how screwed up our ears are. This is why there are loudness buttons on stereos – to boost the lows and highs. You are supposed to turn it on when listening at low volumes. However, most people like extra lows and highs, so they leave the switch on all the time. The main point here is that you should check your mixes at all volume levels. Especially beware of mixing at very low volumes all of the time because you won't be hearing bass and treble as much as you should. Also, whenever you do a fade at the end of a song, the bass and treble will drop out first. Technically, your ears give you the flattest frequency response at around 85 dB.

"Apparent volume" is, therefore, a combination of decibel level, waveform, and the Fletcher/Munson Curve. But relax. Your brain has it all figured out. Most people have no trouble telling whether one sound is louder than another (although most of us need to learn to hear smaller and smaller decibel differences). Your brain quickly calculates all of the parameters and comes up with the apparent volume. All you have to do is listen to the overall apparent energy coming from each sound in the mix. It is apparent volume that you use to set volume relationships in the mix. You don't look at the faders; you listen for the relative volumes.

> The visual representations used in this book show
> the *apparent volume* of the sound, not just the fader level.

Visual 56.
Volume as Front to Back

Compressor/Limiters

Compression and limiting are volume functions; their main purpose is to turn the volume down. They turn down the volume when it gets too loud – that is, when it goes above a certain volume threshold. When the volume is below the threshold, the compressor/limiter does nothing (unless broken or cheap). The difference between compressors and limiters is explained later in the section on "Ratio Settings."

Compressor/Limiter Functions

Compressor/limiters have two main functions (and three other minor ones). The first function is to get a better signal-to-noise ratio, which means less tape hiss. The second function is to stabilize the image of the sound between the speakers, which means more presence.

Better Signal-to-Noise Ratio: Less Hiss

When recording dynamic sounds, with a wide variation from soft to loud, requires turning the volume down so that the loud sounds don't overload and cause distortion. Distortion is also against the law. Get distortion, go to jail. But when you turn the volume down, the soft portions of the sound barely move the needles on the VU meters and you don't get as good of quality.

Therefore, compressors are used while recording to turn down the volume of any peaks, so you can then record the volume louder on the meters and hard drive. By turning down the peaks, you can record the signal hotter, which is better quality sound.

Compressors were originally introduced in the studio when we used tape players. If you recorded at a low volume on the meters you would hear tape noise. If the needles are hardly moving on a tape player, you hear as much tape hiss as you do signal. This condition is known as a bad signal-to-noise ratio and sounds very similar to an ocean: "shhhhhhhhhhhh." By turning down the peaks you could record hotter so the softer portions were loud enough so that you don't hear the tape noise.

When recording digitally, there isn't much noise to worry about. However, if you record too softly the quality of the recording is not as good. Technically, you are recording at a lower bit rate. Therefore, compressor/limiters are also good to use when recording digitally. In addition, limiters can be used to keep very quick sound peaks that you might not even hear from going into the red and distorting.

Note that this function of recording hotter is accomplished if you use the compressor while recording. On a Digital Audio Workstation, this normally means creating bringing in the microphone onto an Auxiliary Track instead of an Audio Track. You put the compressor on the Auxiliary Track and buss it over to an Audio Track. That way the compression is recorded on the audio track.

Stabilizing the Image of Sounds: More Presence

After years of using compression to get rid of hiss, people realized that sounds often appeared more present when compressed. By evening out the volume peaks on a sound, a compressor/limiter stabilizes the image of the sound between the speakers. A sound naturally bounces up and down in volume, as shown by the bouncing VU meter. When a large number of sounds fluctuate naturally, their bouncing up and down can become extremely chaotic – similar to trying to watch twenty-four VU meters at once. A compressor stabilizes, or smoothes out, the movements of sounds that result from these moment-to-moment fluctuations in volume. Once compressed, the sound no longer bounces around much, so the mind can focus on it better. Therefore, the sound seems clearer and more present in a mix.

If you have an instrument that is playing a lot of notes (especially percussion) you just don't have time to hear subtle volume fluctuations. Therefore, any really busy musical parts are compressed more. Also, the busier the mix (the more instruments and the more notes per instrument), the more the sounds in the mix are normally compressed. If you have a 100 piece choir and one person is low in volume or sings something softly you won't hear them at all. By stabilizing the sounds, the entire mix becomes clearer.

Most "acoustic" sounds are compressed, although different engineers disagree about whether to compress live drums or not. Often the kick drum is compressed, and then the entire drum set is compressed with a stereo compressor.

Sounds that are really dynamic (that range from very soft to very loud) are also compressed more. In addition, once a sound has been stabilized, you can then turn up the overall volume and put the whole sound right in your face. This is commonly done in radio and TV commercials to make them sound louder, so that they jump out and grab your attention. This might be annoying in radio and TV commercials, but it's great for a lead guitar or any other special or unique instrument you want extremely present in the mix.

This also works when putting sounds in the background. The problem with low volume sounds is that they can easily be lost (masked by the other sounds) in the mix, especially if the volume of the sound fluctuates very much. Therefore, it is common to stabilize sounds more that are going to be placed low in the mix with compression. They can then be placed extremely low in a mix so even if someone sings or plays a note softly it won't be lost in the mix.

You can also use a compressor/limiter on some sounds as a special effect. Heavy compression or limiting tends to make a sound seem unusually up front – almost as if it is inside your ear.

Finally, the certain styles of music, such as pop, are commonly compressed more.

Note, stabilizing a sound in a mix might also be done during mixdown. However, if you compress a signal enough while recording then you shouldn't have to re-compress during the mix.

Sharper or Slower Attack

Besides less hiss and more presence, a compressor/limiter also makes the attack of a sound sharper. Once you turn down the louder part of a signal, a sound reaches its maximum volume much quicker.

With a shorter and sharper attack, sounds are much tighter, punchier, more distinct, and more precise, which makes them easier to dance to. On the other hand, a higher quality, fast compressor will actually help to remove sharp "spikes" on the attack of a sound – softening the sound. A good compressor can mellow out the sound of a guitar with a sharp edge on the attack.

More Sustain

A compressor/limiter is also used to create more "sustain." This is commonly used on a guitar sound. Just as a compressor is used to turn down the volume peaks to raise a sound above the tape noise, it can also be used to turn down the louder parts of a guitar sound, so the guitar can be raised above the rest of the mix. Sustain is also especially helpful for obtaining the desired screaming feedback (when the guitar is held directly in front of a guitar amp).

Compressors are sometimes used in the same way to create more sustain on tom and cymbal sounds. The sounds seem to last longer before they fade out or are absorbed into the mix. The trade-off is that compressing toms and cymbals will bring their level down, so that you actually hear the bleed more. However, depending on your musical values and the project you're working on, you may want to give this a try.

Less Resonance

A final function of a compressor/limiter is that it evens out resonances in a sound. Resonances create volume peaks!

Resonances occur in two places in instruments: hollow spaces and materials. When a hollow space (like the body of an acoustic guitar) has two parallel walls, it will boost the volume of particular resonant frequencies. Tap on the body of an acoustic guitar in different places, and you can hear the resonant frequencies. Materials (like the neck of a bass guitar) will also resonate at certain frequencies, boosting the volume of those frequencies. Play any guitar, and you will notice that certain notes are louder and more resonant than others because they are activating the resonances in the body of the instrument. A compressor turns down these volume peaks caused by resonance, evening out all of the frequencies in the sound.

Resonances also happen in rooms with parallel walls. The room can also boost the volume of certain notes. Again, the compressor will turn down the volume of these peaks, so you hear all of the harmonics (frequencies) in the sound more evenly – normally giving you a richer and more interesting sound.

Visual 57.
Actual EQ Curve of Resonance
(Silicon Graphics "AMESH"
Spectrum Analysis)

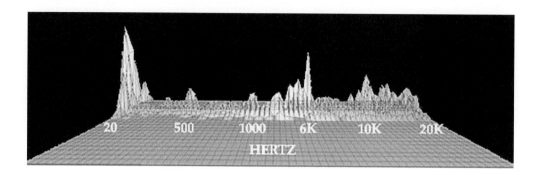

A compressor/limiter evens out the volume of these resonances by turning down the loudest part of a sound, which just happens to be the resonances.

Visual 58.
Resonance Flattened Out

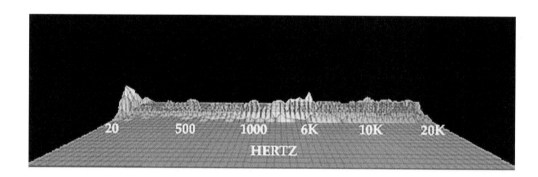

This is why compressor/limiters are so commonly used on resonant instruments such as bass guitar, acoustic guitar, and voice.

Compressor/Limiters: How to Set Them
Most compressor/limiters have two main controls, commonly known as the threshold knob and the ratio knob. On some units the threshold is called "trigger gain," "input," or "compression."

Ratio Settings
The ratio settings control how much (by percentage) the sound volume will be turned down when it goes above the threshold. For example, if a sound is 10 dB above the threshold and the ratio is set to 2:1, it will be turned down 5 dB. If a sound is 30 dB above the threshold, it will be turned down 15 dB. Ratio settings normally range from 2:1 to ∞:1 (infinity to one).

Visuals are especially effective in explaining the functions of the threshold and ratio knobs on compressor/limiters. If volume is shown as a function of front to back, the sphere will bounce back and forth based on the VU meter. It will then come out front and slam into the threshold.

Visual 59.
Sound Smashing into
Threshold on
Compressor/Limiter

The difference between a limiter and a compressor is that a limiter stops the volume from getting any louder than the threshold. The problem is that when a sound is steadily rising in volume then suddenly stops cold at the threshold, it doesn't sound natural to our ear. It sounds squashed. A compressor on the other hand, allows the volume to get a bit louder than the threshold based on a ratio, or percentage. If we set the ratio to 2:1, it will go this far:

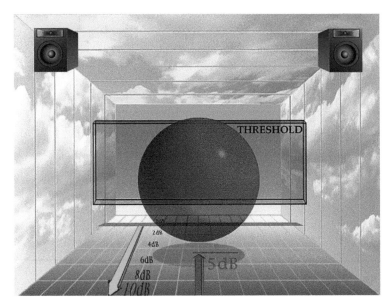

Visual 60.
2:1 Ratio on
Compressor/Limiter

A good starting point is the ratio of 4:1; this will still turn the volume down, but won't squash it. You can set the ratio wherever you like, but most people just starting out can't hear the difference between ratio settings very well. Until you can, 4:1 is a good place to start.

> Some people prefer a 3:1 ratio. On some units you can hear the squashed sound sooner, so people use a lower ratio setting. On some really nice units, you can set the ratio really high before the squash sound becomes noticeable.

Threshold Settings

As the threshold is lowered on a compressor/limiter, the volume, or gain, of the sound is reduced. The compressor/limiter meters or LEDs labeled "gain reduction" will then bounce backward (or down), showing the exact amount of volume reduction at each moment.

When adjusting the threshold, don't look at the threshold knob; rather, watch the gain reduction meters because the threshold directly affects the amount of gain reduction. Turn the threshold knob until you get a maximum of 6 dB gain reduction. If you set the threshold lower so you get more gain reduction, it will sound like it is squashed.

However, for some instruments, the threshold is commonly set to provide a maximum of 10 dB of gain reduction. As mentioned before, the following are compressed more by using 10 dB of gain reduction:

- Dynamic sounds or instruments
- Sounds or instruments that are mixed right up front or way in the background using volume.
- Busy sounds
- Busy mixes

Again, once you can hear the nuances of various compression settings, you can set ratio and threshold the way you want for the style of music, the song, and the sound itself. Until then, use the following settings: set the ratio at 4:1 and the threshold for 6 dB of gain reduction.

Mastering

Overall compression and limiting on the whole mix is one part of mastering.

Compression is used to even out the overall mix. Limiting is used in order to get the signal as loud as possible within the volume limits of a CD.

The compressor is commonly set at a ratio between 2:1 and 4:1, depending on the smoothness of the unit being used. Any "squashed" sound is especially annoying on an entire mix. The threshold is set so that you are getting a maximum of 3–6 dB of gain reduction on the gain reduction meters. For very sparse acoustic music, I will set the threshold for only a maximum of 1 or 2 dB. Again, any more than that, and you can normally here the sound being squashed.

The mastering limiter is set to a ceiling of around negative 0.3 dB, which means that it will be as loud as the CD can handle. The threshold is set to get a maximum of 3 dB (often only 1 or 2 dB) on the gain reduction meters. Be especially careful to not limit too much because it is very noticeable and especially annoying. Most mastering limiters turn the volume up, automatically compensating for the amount of gain reduction you are getting. Therefore, you normally don't have to set the output volume of the limiter.

Noise Gates

Operating similarly to a compressor/limiter, a noise gate turns the volume down (therefore, compressor/limiters and noise gates are often packaged together in one box). The difference is that a compressor/limiter turns the volume down above the threshold, while a noise gate drops the volume when falls below the threshold. However, since the volume is being turned down on a sound that is already low in volume, normally a noise gate will turn off the sound completely.

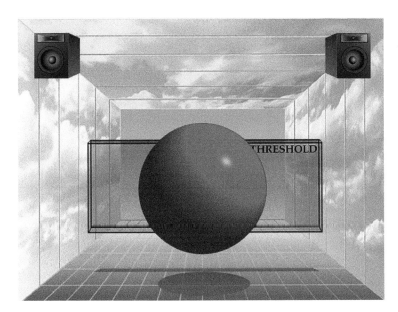

Visual 61.
Sound Fading Out
Past Threshold on
Noise Gate

Noise gates have three main functions: to get rid of noise, to get rid of bleed, and to shorten the duration of a sound.

Noise Eradication

The first function of a noise gate is to get rid of noise, hiss, or anything annoying that is low in volume. Noise gates only get rid of background noises when a sound is not playing. Noise gates don't get rid of noises while the main signal is present; however, you normally can't hear the noise when the sound is playing.

One function of a noise gate is to get rid of amp noise when a guitar is not playing. Say you have a guitar amp set on 11 with lots of distortion. When not playing, the amp makes a huge "cushhhhhh" sound (when the guitar is playing, you don't hear the amp noise because the guitar is so incredibly loud). You set the noise gate by having the guitar player hit a note and sustain that note until it fades and the noise of the amp takes over. Set the threshold of the noise gate so that when the volume fades enough to hear the amp noise, it gets cut off. Whenever the guitar player is not playing, you now hear silence.

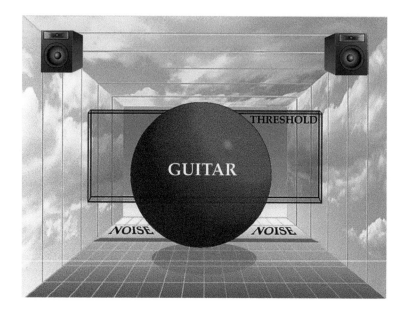

Visual 62.
Noise Gate: Threshold
Set Between Guitar
Sound and Noise

It is important not to chop off any of the guitar sound. All it takes is for the musician to play a soft note, and the noise gate will chop the sound right off. Noise gates can also be used to get rid of noise from tape hiss, cheap effects units, dogs, kids, and crickets.

Bleed Eradication
Another common use of a noise gate is to remove the bleed from other instruments in the room. When a mic is on an instrument, the sound of that instrument will be loudest in the microphone. Therefore, it is easy to set the threshold of a noise gate between the sound and the bleed, so that the bleed gets turned off.

Visual 63.
Noise Gate: Threshold
Set Between Sound and
Bleed

The obvious advantage of isolating a sound like this is that you have more individual control over volume, equalization, panning, and effects. Once a sound is isolated with a noise gate, any changes you make with a sound manipulator will only change the one sound you are working on. Gates can be especially effective on drums to isolate each drum. This is especially important on a snare if you are going to put a lot of reverb on it. Without the gate, you end up with reverb on the hi-hat as well. Another advantage of isolation is that it helps to eliminate phase cancellation (we'll discuss this more later).

But most importantly, by removing the bleed, you will then hear the sound in only one microphone. This has the effect of putting the instrument in one precise spot between the speakers, instead of being spread in stereo. For example, consider the miking of a hi-hat cymbal. Besides being picked up by the hi-hat mic, the hi-hat is also being picked up by the snare drum mic. If the hi-hat mic is panned to one side and the snare mic (with the hi-hat bleed) is panned to the center, the hi-hat then appears to be spread in stereo between the speakers. It is no longer clear and distinct at a single spot in the mix (technically, phase cancellation also occurs, which also makes it less clear). Putting a noise gate on the snare mic turns off the hi-hat when the snare is not playing. The isolated image of the hi-hat, in its own mic, will now appear to be crystal clear and precisely defined wherever the hi-hat mic is placed in the mix.

Shortening the Duration

You can also use a noise gate to shorten the duration of a sound. The noise gate will cut off both the attack and release of a sound because these are commonly the softest parts of the sound. This can be quite an interesting and unusual effect.

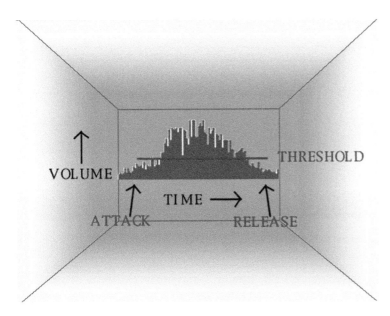

Visual 64.
Noise Gate Cutting off Attack and Release of Sound

A noise gate can also be put on reverb to chop off the release, resulting in the well-known effect referred to as *gated reverb*. Normally, we just pull up the gated reverb setting in the reverb effect unit.

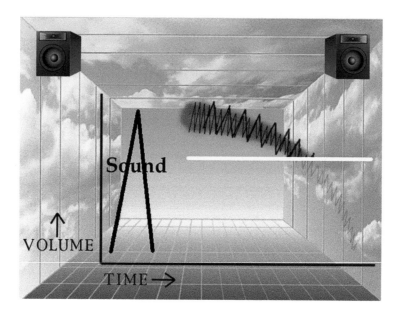

Visual 65.
Envelope (Change in Volume over Time) of Gated Reverb

Visually, when volume is shown as front to back and the volume is less than the threshold setting, the sound will disappear. If the low volume sound is noise, bleed, or the attack and release of a sound, it gets cut off.

Section B: Equalizers

EQ is a change in the volume of a particular frequency of a sound, similar to the bass and treble tone controls on a stereo. It can be one of the most difficult aspects of recording and mixing to master because there is such a large number of frequencies – from 20 to 20,000 Hz. Different sounds are equalized differently depending on the type of music, the song, and even the people you are working with.

First you must learn all the frequencies or pitches by name. Then, you will see how boosting or cutting a certain frequency affects different instruments in different ways.

Frequency (Pitch)

The primary difference between frequency and pitch is
frequencies are labeled with numbers and
pitches are labeled with letters.
Also frequencies are way more detailed than pitches.
There are many frequencies that are the same note.
Also frequencies are way more detailed than pitches.
There are many frequencies that are all the same pitch.

Octave 0		Octave 1		Octave 2		Octave 3	
A	27.500	C	32.703	C	65.406	C	130.81
A♭	29.135	C♯	34.648	C♯	69.269	C♯	138.59
B	30.868	D	36.708	D	73.416	D	146.83
		E♭	38.891	E♭	77.782	E♭	155.36
		E	41.203	E	82.407	E	164.81
		F	43.654	F	87.307	F	174.61
		F♯	46.249	F♯	92.499	F♯	185.00
		G	48.999	G	97.999	G	196
		A♭	51.913	A♭	103.83	A♭	207.65
		A	55.000	A	110.00	A	220.00
		B♭	58.270	B♭	116.34	B♭	233.08
		B	61.735	B	123.47	B	246.94

Octave 4		Octave 5		Octave 6		Octave 7	
C	261.18	C	523.25	C	1046.5	C	2093.0
C♯	277.18	C♯	554.37	C♯	1108.7	C♯	2217.5
D	293.66	D	587.33	D	1174.7	D	2349.3
E♭	311.13	E♭	622.23	E♭	1244.50	E♭	2489.0
E	329.63	E	659.26	E	1318.5	E	2637.0
F	349.23	F	698.46	F	1396.9	F	2793.0
F♯	369.99	F♯	739.99	F♯	1480.0	F♯	2960.0
G	392.00	G	783.99	G	1568.0	G	3136.0
A♭	415.30	A♭	830.61	A♭	1661.2	A♭	3322.4
A	440.00	A	880.00	A	1760	A	3520.0
B♭	466.16	B♭	932.33	B♭	1864.7	B♭	3729.3
B	493.88	B	987.77	B	1975.5	B	3951.1
						C	4186.0

Chart 2. Frequencies Corresponding to Pitches

Frequency Ranges

Learning all of the frequencies by name is easier than you might think because we already know all the frequencies by heart. Our entire system has been perceiving frequencies since the day we were born (and even before). We are all professional listeners with years of experience at differentiating between frequencies. In fact, we are frequency beings – every part of our body from cell to soul is a frequency.

Chart 3.
Six Frequency
Ranges

HI HIGHS   >8000   Hertz
HIGHS   5000-8000 Hertz
MIDRANGE 800-5000   Hertz
LOW MIDS   100-800   Hertz
BASS   40-100   Hertz
LOW BASS   <40   Hertz

When you learn the names of the frequencies, you can begin to remember what boosting or cutting each frequency does to each instrument. It also helps with more precise communication when discussing what to do with an equalizer. In order to make all the frequencies in the spectrum easier to remember, they can be divided into six ranges. There is no commonly accepted framework for dividing the frequencies into ranges (every book seems to divide the frequencies a little differently). Here is an average of the way, frequency ranges are divided in most books on the subject.

Sub Bass (Low Bass): Less Than 40 Hz

This range is commonly found in rap booms and the low bass of kick drums and bass guitars. It is difficult for many people to discern pitch easily at this range. These frequencies are often accentuated in movies for earthquakes, rumbles, and explosions.

Because records could not reproduce frequencies this low, we used to roll off this range as a matter of course. The grooves on a record have to be wider for bass frequencies than high frequencies. If you didn't get rid of the low bass frequencies, you wouldn't be able to get the normal 23 minutes on one side of the LP. (This is also why you can put more low bass on a 12-inch single record.) Of course, this is not an issue with CDs.

Bass: 40–100 Hz

This is the approximate range boosted when you turn up the bass tone control on a stereo. 40–60 Hz is the nice low end that we love. Sounds already start to get a bit muddy as low as 70 Hz.

Low Mids – Oohzone: 100–800 Hz (Primarily in the 100–400 Hz Range)

This range is called the *oohzone* because, if it is boosted too much, it makes most people go "uuu" (more like "ewww" with a negative tone and your nose up). In addition, the resonant frequency of your chest cavity is within this range. That means your chest is boosting the volume of some frequencies within this range. You can feel it if you put your hand on your chest and hum a low-frequency "uuu." (Did you do it?)

I often refer to this range as the "muddy" range. When frequencies are boosted too much in this range, they sound extremely muddy and unclear and can even cause extreme fatigue when not evened out. You'll also find that everybody in the room starts getting a bit irritated. On the other hand, this range is also thought of as "warmth." If you take it out too much, your mix might sound thin and anemic.

Midranges: 800–5,000 Hz

This is the "presence" range. If it is weak or missing, sounds normally seem more distant. We are extremely hypersensitive to this frequency. This is where we live – our language is centered here. Boosting a frequency 1 dB in this range is like boosting 3 dB in any other frequency range. Therefore, you want to do approximately 1/3 less boosting or cutting in this range. It is critical to be careful when boosting or cutting any frequencies here. This is doubly true on vocals because we are so tuned into what a natural vocal should sound like. It's funny – you may not know what a natural guitar sound should sound like (at this point), but we all know what a natural vocal should sound like. Therefore, you might do 1/6 of the normal EQ when working on vocals!

The telephone is centered around 3,000 Hz because we can still understand someone when only this range is present. This makes it really tempting to boost this range to make a vocal more intelligible. It will work; however, it also makes it sound unnatural (like a megaphone). And as mentioned, we are overly sensitive to vocals and this frequency range.

Other notable frequencies in this range include 1,000 Hz, which is the frequency of TV stations' test tones when they go off the air. The chainsaw frequency is around 3,000–4,000 Hz, and it is the most irritating frequency by far. It is also the frequency of fingernails on a chalkboard.

Highs: 5,000–8,000 Hz

This range, the one boosted when you turn up the treble tone control on a stereo, is often boosted in make things sound brighter and more present.

Hi-Highs: More Than 8,000 Hz

This is where you find cymbals and higher harmonics of sounds. Boosting this range a little on certain instruments can make the recording sound like a higher quality recording, but too much can make it irritating. It is also the range where most hiss exists. By the way, that extremely high frequency that old televisions emit is 15,700 Hz.

The Complexities of Frequencies: The Harmonic Structure of Sound

Most sounds are made up of a combination of pure tones, called harmonics or overtones. When you hear an instrument play a particular pitch, there are actually many other notes hidden in that sound. For example, if you just make the sound "Ah," there are over 100 different harmonics in that one sound. Also, all of these harmonics are actually notes. However, our brain can't handle it so we just tune into the one "fundamental" note at the bottom or all of the other frequencies. Therefore, just about all sounds are made up of many frequencies.

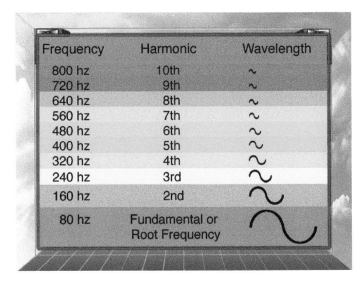

Frequency	Harmonic	Wavelength
800 hz	10th	~
720 hz	9th	~
640 hz	8th	~
560 hz	7th	~
480 hz	6th	~
400 hz	5th	~
320 hz	4th	~
240 hz	3rd	~
160 hz	2nd	~
80 hz	Fundamental or Root Frequency	~

Visual 66.
Pitches of Harmonics

For example, here is the harmonic structure of the note "A" on an acoustic guitar. Just look at all the notes present when you play what most people think of as one note:

Frequency	Harmonic	Pitch
4400	10th	C#
3960	9th	B
3520	8th	A (3rd Octave)
3080	7th	G 1/4 flat
2640	6th	E
2200	5th	C#
1760	4th	A (2nd Octave)
1230	3rd	E
880	2nd	A
440	Fundamental or Root Frequency	A

Visual 67.
Harmonic Structure of
Note "A" as on Guitar

It is the particular harmonics present in a certain sound that account for the differences in timbre. The term timbre refers to different sound qualities or tonalities, such as guitar versus piano or then tonality of different voices, as well as the differences in the sound quality of particular instruments (e.g., the difference between a Martin and a Gibson guitar).

There are two interesting things about harmonics. First, each harmonic found in a sound's timbre is a pure tone. A pure tone is the sound of a tuning fork or tone generator. It has no harmonics at all. The most amazing thing is that just about all sounds are made up of a combination of these pure tones. This means that even a screaming electric guitar sound is made of many pure tones.

So how do you get an edgy sound from a bunch of pure tones? Well, certain combinations of harmonics create a dissonant chord. These are the odd numbered harmonics. If you play a bunch of notes that are not in key or in tune, they will sound quite edgy and irritating, like the sound of Axl Rose or Tiny Tim's voice (may he rest in peace). On the other hand, certain combinations of harmonics will create a chord that sounds good. These are the even-numbered harmonics. If the pitches of the harmonics combine to create a nice chord, the sound will be nice and round (like the voices of Barry White or Norah Jones).

Whether an instrument puts out odd or even harmonics is based on the construction of the instrument and how the sound is produced. Odd harmonics are created by rough textures (throat of Janis Joplin), asymmetrical shapes, chaotic movement (hitting an object or the movement of a reed), metal (steel strings or Tibetan bowls), and distortion (hard rock guitar sound). Whereas even harmonics are created by smooth surfaces and symmetrical shapes, nylon strings (harp or classical guitar), or wood (wood or tongue drums).

Also, when an instrument is played really loudly, it will put out more odd harmonics. When someone screams, the edge to the voice is the odd harmonics. The sound of a drum being hit really hard contains more odd harmonics. A really loud amp puts out more odd harmonics because both the tubes (or other electronics) and the speaker are being stretched and pushed to their limits. Therefore, we can recognize loud sounds even when they are placed really low in the mix *because our ear still hears the odd harmonics*. For example, a screaming guitar or vocalist will still sound loud, even if each is placed so low in the mix that you can hardly hear them. This is an important to remember if you are wanting to make a mix sound loud. You simply use sounds that are played loudly. Regardless of how loud they are in the mix, they will still sound loud.

A sound that has no harmonics is called "pure tone." These are mostly tuning forks and tone generators. Sounds that have very few harmonics are called "pure sounds." These include flutes and crystal singing bowls. Sounds with a large number of harmonics are called rich or complex. These include the violin, large instruments such as the piano, and the sound with the most harmonics of all – the voice.

Here is where it all matters:

> **When you raise or lower**
>
> **the volume of a certain frequency with equalization,**
>
> **you are actually raising or lowering**
>
> **the volume of a particular harmonic in the sound.**

If a sound has no harmonics an equalizer is useless. You can only turn up or down the whole sound because it is only one frequency. Likewise, with a flute, there is very little EQ'ing that can be done. On the other hand, you can EQ the voice all day long because there are 100's of harmonics to play with.

Commonly what we are doing is turning down odd harmonics to make sounds more mellow or palatable. Or, turning up the odd harmonics to make the sound cut through even more.

Tubes in microphones or mic preamps actually create even harmonics, which make a sound warmer.

The important point here is because every sound has its own harmonic structure, every instrument sound responds to equalization differently.

Types of Equalizers

There are three main types of equalizers found in the recording studio: graphics, parametrics, and rolloffs (highpass and lowpass filters).

Graphics

Each frequency can be turned up or down by using the volume sliders on a graphic equalizer. The volume controls on an equalizer are called bands. There are different kinds of graphic equalizers that can divide frequencies from five bands up to thirty-one bands. 5–10 band graphic equalizers are commonly found in car stereos. Thirty-one band graphics (which will change the volume at thirty-one different frequencies) are common in recording studios and live sound reinforcement on the entire sound system. They are commonly used to compensate for resonances in speakers or a room.

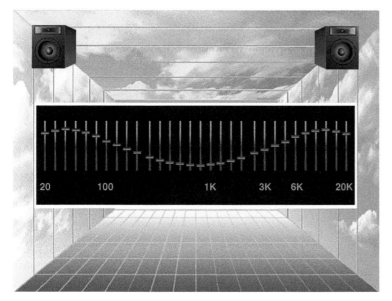

Visual 68.
31-Band Graphic EQ

The primary advantage of a graphic equalizer is that you can make changes in volume at a number of different frequencies. Graphic EQs got their name from the visual display that's easy to read for reference. (However, these days, you get a much nicer display on a digital parametric EQ.) Also, since the frequencies are mapped out visually from left to right, it is easy to find and manipulate the volume of any particular frequency.

Many people don't realize that when you turn up a particular frequency on a graphic, you are actually turning up a range of frequencies preset by the manufacturer. For example, if you turn up 1,000 Hz, you might actually be turning up a frequency range from around 300 to 5,000 Hz.

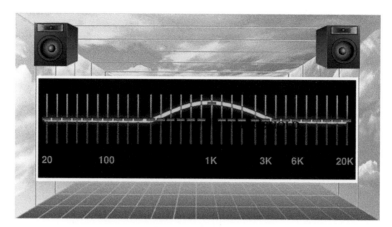

Visual 69.
Wide Bandwidth on
Graphic EQ

This range of frequencies is called the bandwidth and is preset by the manufacturer. You have no control over the bandwidth on a graphic. Generally, the more bands (or volume controls) there are, the thinner the bandwidth. Therefore, a Thirty-one-band graphic EQ will have a more precise frequency range for each slider than a five-band graphic. If you turn up the volume of 1,000 Hz on a five-band graphic, you could be turning up from 100 to 10,000 Hz.

Using the book's visual framework, frequency is shown as a function of up and down, so highs to lows are shown in a graphic representation like this:

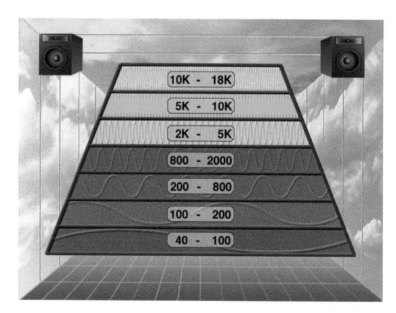

Visual 70.
Virtual Mixer
Graphic EQ

The volume of a particular frequency is shown as the brightness in that band. For example, if you turned up the highs around 1,000 Hz, you would see it get brighter in that frequency range, like this:

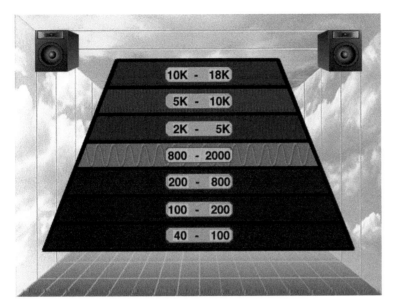

Visual 71.
1,000 Hz Boost

Showing an equalizer this way is useful when sounds overlap in a mix. You can then see the density of the frequencies at any spot in the mix.

Visual 72.
Virtual Mixer
Spectrum Analysis

Parametrics

Engineers want to be able to control the range of frequencies, or bandwidth, they are turning up or down. With a parametric, the bandwidth knob gives you control over the width of the frequency range being manipulated. The knob is usually called "Q" because the word "bandwidth" won't fit on the knob ("Q" stands for "quality," which is an electrical term for bandwidth). A thin bandwidth is normally labeled with a peak, whereas a wide bandwidth is often labeled with a wider hump. Sometimes ranges of musical octaves are used to define the bandwidth – for example, from 0.3 octaves to 3 octaves wide. Sometimes a scale of 1–10 or 10–1 (it's not standardized in DAW's) is used.

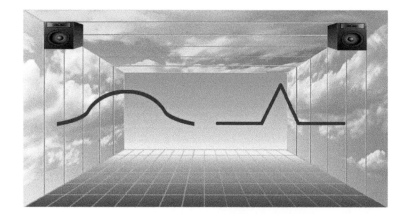

Visual 73.
Wide and Narrow
Bandwidth Symbols on
a Parametric EQ

Using the visual framework, the bandwidth can be shown with narrower or wider bands of color.

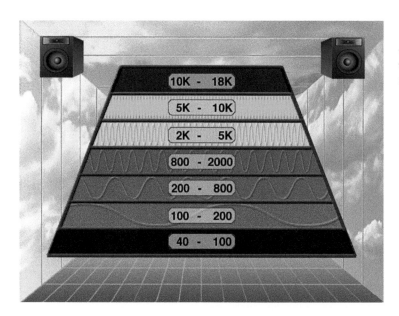

Visual 74.
Wide Bandwidth of
Frequencies Boosted

On a graphic equalizer, you select the frequency by moving your arm left to right to place your hand on the correct volume slider. On a parametric EQ, you select the frequency by turning the "frequency sweep" knob with two fingers. A separate volume knob is then used to turn the chosen frequency up or down.

Paragraphics
Some less expensive consoles have equalizers with frequency sweep knobs but do not have bandwidth knobs. This type of equalizer is commonly referred to as semi-parametric, quasi-parametric, or para-graphic. Be careful, though; these days some manufacturers and certain salespeople are now using the term "parametric" to refer to a paragraphic or semi-parametric, even though it has no bandwidth control.

Rolloffs

A rolloff EQ turns down the volume of low or high frequencies. They are commonly found on consoles as highpass and lowpass filters. Larger consoles often have sweepable or variable rolloff knobs so that more of the lows or highs are rolled off. Smaller consoles often have only a button that rolls off a preset amount of lows or highs. A highpass filter rolls off the low frequencies but does nothing to the highs; it passes them.

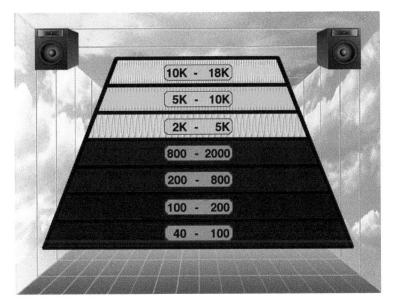

Visual 75.
Highpass (Low-Cut)
Filter

Highpass filters are especially helpful in getting rid of low-frequency sounds, such as trains, planes, trucks, air conditioners, earthquakes, bleed from bass guitars or kick drums, and serious foot stomping.

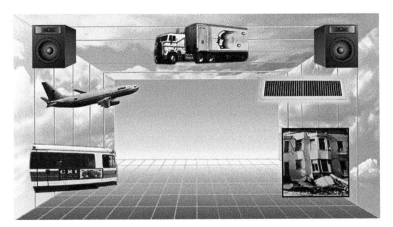

Visual 76.
Things That Rumble

Highpass filters can be found on microphones and smaller mixing consoles as switches that simply roll off the lows when the switch is engaged.

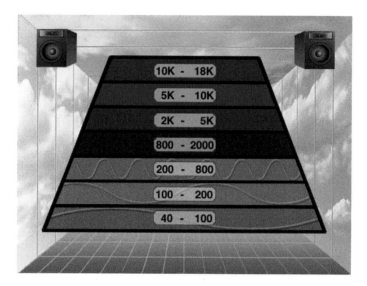

Visual 77.
Lowpass (High-Cut) Filter

A lowpass filter rolls off the high frequencies and is especially helpful in getting rid of hiss on sounds that don't have a lot of highs in them, such as a bass guitar. Lowpass filters are also used to roll off the high-frequency attack (click) of a kick drum in order to make it sound more like the classic rap kick drum sound.

Using Equalizers

When to Equalize

There are five occasions when you might equalize a sound in a recording session. Sooner or later, you will probably get it right. It's always better sooner than later.

1. *In solo before recording.* Each sound is equalized individually, one by one, before recording.
2. *In the mix before recording.* While the whole band is playing double-check the EQ of each sound relative to all the other sounds in the mix.
3. *In solo during mixdown.* Each sound is again equalized individually before building the mix.
4. *In the mix during mixdown.* Finishing touches are done while listening to the whole mix at once. The most important step, obviously.
5. *During mastering.* Finally, a bit of EQ is occasionally done during the mastering process. This overall EQ for the entire mix might not be necessary if a good job was done in the first place.

Equalizing in Solo Before Recording onto the Multitrack

The first step in the recording process is to equalize each sound individually. Most engineers start with the drums. Have the drummer play the kick drum by itself over and over. Go through each drum sound and every other instrument one by one and EQ them individually.

There used to be a school of thought that said you should not EQ a sound going to the multitrack. This idea was the result of inexperienced engineers who had screwed up the EQ on the way to the multitrack. It is then very difficult to get a sound back to normal and still be able to make it sound great during mixdown. Therefore, it is important that you EQ the sound correctly onto the multitrack in the first place. And, as you will see, it's not that difficult. The school of thought these days is to definitely EQ on the way to the hard drive. In fact, professional engineers will usually try to get everything to sound like a CD on the way to the multitrack. There are some very important advantages for doing this.

First, if you get the EQ right before you record everyone is happier, gets higher and plays better simply because it sounds better in the headphones while recording. The energy in the room is completely different. Neurons of creativity often begin firing off everywhere because it sounds so incredible. A great mix turns people on creatively. This becomes a huge deal when overdubs take a few weeks. These days, most bands, especially those who have worked in major studios, expect you to get it sounding as close to a CD as possible on the multitrack.

Second, if you EQ during recording, you can save processing power by not having to pull up an equalizer during the mixing process. If you really need to also EQ in the mix don't hesitate.

Third, when you get to the final mix, it is really nice to have the EQ already sounding close to perfect. Instead of spending your time getting the EQ in the ballpark, you can now focus on fine, subtle details that bring out some serious magic in the sound.

The question is, "What is good sounding?" In the beginning, "good" meant "natural." The trick was to go out into the studio, listen to the sound, and then use the EQ to get it to sound the same way in the speakers in the control room. This is still a good technique. However, as I'll discuss in detail later, natural isn't natural like it used to be. Things are brighter, bassier, and more present than in the old days. And now, occasionally natural is not the goal at all. Sometimes you can be creative and produce an "interesting" EQ. Normally, the goal is to make a sound natural. Typically, this means making sure the sound is not too muddy, too irritating, too dull, and has enough low bass.

The professional engineer gets to the point where he or she can guess what a sound should sound like in solo in order for it to sound right in the mix. This will come with experience as you get the natural sound of instruments in your head. It is also helpful if you have heard the song before with all the instruments together.

By the way, you often don't need to EQ synthesizer sounds very much, if at all. Most of this is referring to recording live instruments.

Equalizing in the Mix Before Recording Onto the Multitrack

The next time to check the EQ of each sound is when the entire band plays all at once. You can then hear the EQ of each instrument relative to every other instrument in the mix. Just like colors, the EQ of one sound can change dramatically when it is next to another sound. For example, perhaps the guitar doesn't sound bright enough compared to the brightness of the snare.

It is important (in fact, critical) to check the relative EQ of each sound relative to all the others at three frequency ranges: bass, midrange, and highs.

First, scan the high frequencies by listening to how bright each sound is compared to the overall brightness of all the other sounds. Don't solo each sound, just focus your mind on each sound one by one while the whole mix is playing. Make sure all of them are as bright as you want them. Most often, you want them to have a similar amount of brightness, but sometimes you might want some sounds to be brighter or duller than others. If you don't set it, the wind is setting them for you. Occasionally, the wind gets it together and you get lucky. But normally, the wind sucks at mixing. Rather . . . the wind *blows*!

Visual 78.
Song with High Frequencies Highlighted

Second, scan the midrange frequencies focus on each sound one by one, listening for the relative volume of these frequencies across all the instruments. Midrange frequencies seem to stick out when boosted too much. When there is not enough midrange, a sound will seem distant, unclear, and muted. Make sure that all instruments have the exact amount of midrange frequencies that you want. Normally, you want sounds to have a similar amount of midrange frequencies, but sometimes you may want some sounds to stick out more so they grab your attention. Occasionally you want sounds to be more distant, in order to make other sounds seem more present, relatively. Again, if you don't set the midrange the way you want it, the wind will . . . then, good luck.

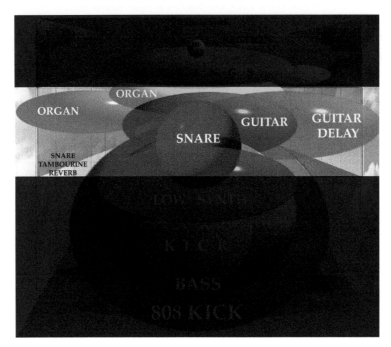

Visual 79.
Song with Midrange Frequencies Highlighted

Third, scan the bass frequencies, checking for the relative volume of bass in each sound in the bass range. For example, check the relative amount of bass frequencies present in the kick drum compared to the bass guitar. This frequency range is the most commonly missed when mixing an album or project. Listen and make sure that it is the way you want it to be. Because if you don't set it the way you want, it's up to the wind. And . . . *the wind blows*!

Homework for the Rest of Your Life – Due (Do) Every Day

Have you ever paid attention to the difference in the amount of bass EQ on the bass guitar compared to the bass EQ on the kick drum? Which should have more? The truth is that it depends on the style of music and the song itself (and sometimes the opinions of the bass player and drummer). For example, in reggae and blues, the bass often has more low end. The kick (especially the 808 boom) in rap often has the most low end. Start checking it out in songs from now on and see what others are doing. And don't forget to ask yourself if you like what they did. Very soon, you will develop your own values as to how much bass you like on the kick versus the bass for different styles of music and songs.

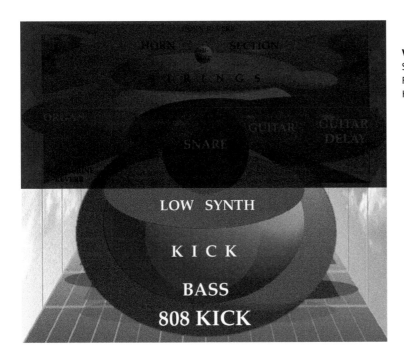

Visual 80.
Song with Low
Frequencies
Highlighted

It is critical that you check the relative EQ of each instrument in the mix at each frequency range. The amount of time you spend doing this often depends on the band. Some bands expect to be recording within a few hours of the time they arrive and have very little patience (or money). Other bands spend days getting the right sound and EQ before recording to the multitrack. It is a good idea to set up the band the day before the session, get all the sounds EQ'd, get a headphone mix, then go home. The next day everything is set to go, and everyone is fresh.

It is also a good idea to talk with the band beforehand and let them know that you will be spending a good amount of time working on the sounds at the beginning of the session. If the band knows what is going to happen, they won't get frustrated while waiting to actually begin recording. Hopefully, they will appreciate the fact that you want it to sound as good as possible.

Equalizing in Solo During Mixdown

When you go to mixdown a song, the first step is to EQ each of the sounds individually. If you did your job well during the recording session (and had enough time), you might have little or no EQ'ing to do. However, often you will need to EQ the sound again because you know what the band is going for and you have a new perspective – a fresh ear. You also have a major advantage that you didn't have when you began the recording session: you know what the whole song sounds like with all of the instruments playing together. Now you can set the EQ of each sound with the final mix in mind.

People often wonder why things don't seem to sound the same when you come back to do the mixdown. It is easy to think that you have something sounding right simply because you have made it sound *so much better* by EQ'ing it while recording. When a band first comes in, often you listen to the sounds (and how bad they might sound), then you EQ them and you're happy because you've made the sound great compared to the original sound. The problem is that you should be EQ'ing the sound based on the real world sound of current CDs. You might have made it sound light years better, but it needed more EQ'ing . . . to sound like a CD. When you come back in to mix the song down, you have been listening to the radio or CDs in the real world, and your ears are attuned to the current natural sound. When you put on the multitrack, you automatically compare your recording to the real world and realize it didn't sound as good as you thought when you did the recording session.

Equalizing in the Mix During Mixdown

The way the EQ sounds in the mix during the mixdown is the true test. Again, you should scan each frequency range – highs, midranges, and lows – and set them the way you want. If it is already pretty good, you can now work on fine-tuning the relative EQs. This is where you do the magical stuff. For example, you might add a tiny bit of 12,000 Hz on some of the high-frequency sounds to make the overall mix sparkle. Or you might consider making a guitar solo track a little brighter and edgier, so that it really cuts.

At this point, besides just setting a static EQ, occasionally it might be appropriately creative to actually turn the EQ knobs while the song is playing. Perhaps you might EQ an instrument differently for various sections of the song. Or, to really be creative, you could change the EQ in the middle of a section of a song. It is also interesting to EQ a sound so that it seems to be coming out of a telephone. More on this later.

Equalizing the Entire Mix During Mastering

There are two main types of EQ done during mastering. First, minor repairs can be done if the overall EQ doesn't sound quite right, but if the problem is very bad the entire song must be remixed. The goal is to match the overall EQ to the current established overall EQ for the particular style of music. It is quite common to adjust the amount of overall bass or treble slightly. Perhaps you might need to cut or boost some midrange ever so slightly.

Second, the overall EQ can be adjusted to make the overall bass, midrange, and treble more similar from song to song. Again, if the difference is too great, it might require remixing the songs. There is only so much that can be done in mastering with EQ since all the sounds are no longer separate on their own tracks.

Using an Equalizer: Step-by-Step Instructions

Here is a really helpful process for EQ'ing sounds.

There are two overall general steps in the EQ process.

The first step is to find the frequency that needs to be boosted or cut in volume. You do this by boosting one of the EQ volume controls (bands) all the way. Then you sweep the frequency back and forth to find the exact center frequency. If you are working on a frequency that you want to turn down, then you are looking for the worst possible sound. If you are working on a frequency that you want to turn up, then you are looking for the nicest sound possible.

The second step is to then turn up or down the volume of the frequency that you found.

The first step is outlined in more detail in steps 1–5 below. The second step is outlined in more detail in steps 6–8 below.

Here is how this step-by-step process works. First, listen to "Too Much Mudd." Then if there is a problem with that frequency range, you then go through steps 2–8 and fix it. Then you move on to the next frequency range, "Not Enough Lows." If a range is fine, don't do anything. If it ain't broke . . . don't fix it.

1. Listen

The most common mistake made by an inexperienced engineer is to begin turning the EQ knobs before listening. Don't touch the knobs until you know what you want to do! Listen to see if anything is wrong with the sound first, and if it ain't broke, don't fix it.

There are four main things to listen for in a sound. These four areas cover about 95% of all EQ'ing done. The other 5% is when you there might be some really unusual frequencies in a sound, or when there are problems with the equipment or instrument.

Check each sound for the following.

a. Too much mudd or low mids – typically between 100 and 400 Hz.
b. Not enough lows – typically between 40 and 60 Hz.
c. Too much midrange – irritating or "honky" frequencies from 800 to 5,000 Hz.
d. Not bright enough – 5,000–8,000 Hz.

Too Much Mudd or Low Mids – Typically Between 100 and 400 Hz

Check each instrument for muddiness. Always start with muddiness because when it is taken out the sound will often sound brighter. Kick drums almost always need to have the muddiness cut (unless it is a rap or hip hop kick). Other potentially muddy instruments include toms, bass guitar, piano, acoustic guitar, and harp (or any instrument with bass in it). Muddiness is normally between 100 and 400 Hz, however occasionally there might be some work to be done from 400 to 800 Hz. If you cut the muddiness too much, the instrument will sound thin because this mudd also contributes to the body of most sounds. When cutting muddy frequencies, always make sure that you haven't lost your bottom end: the low lows. You might compensate by boosting the lows around 40–60 Hz.

BOOST LOWS (40–100 HZ)

It is common to add a little extra boost to the kick drum or bass guitar depending on the style of music and song. However, do be careful because it is easy to overdo it. Remember that most people boost the bass on their stereos. You need to make it more flat in the studio. Also, in music other than hip hop and dance music, a kick drum with too much bass is the #1 culprit of a muddy mix. Often it is best to simply boost the volume of the overall kick drum by raising the fader instead of boosting the low frequencies.

CUT IRRITATION (800–5,000 HZ)

Cut any excessively irritating or honky frequencies occurring in the midrange from 800 to 5,000 Hz. Vocals, electric guitars, and cymbals (including hi-hat) sometimes need frequencies cut in the midrange. Depending on the type of music (and the particular snare drum used), snares occasionally need a lower frequency "whack" cut also. The best way to detect an irritating frequency is to turn the entire sound up very loud. If you and the people in the room are cringing on the floor, then it's irritating. Remember, we are hypersensitive to the midrange, so never boost or cut the midrange too much. Also, make sure you haven't made the sound too dull. At that point, you might compensate by boosting the highs around 5,000–8,000 Hz.

BOOST HIGHS (5,000–8,000 HZ)

Boosting highs on instruments that sound dull, like the snare, is largely dependent on the style of music. Pop, R&B, dance, and certain types of rock and roll require more crispness than other styles. Country, middle-of-the-road, and folk music might not need as much boost in this range, so they sound more natural.

2. Pull up the Equalizer and Reset to "0"

Pull up the EQ you want to use on the channel. Often I will copy an EQ from another channel (normally you drag it to the new channel while holding down Option on a Mac or Alt on a PC). When copying the EQ from another channel, you might need to reset the volume controls on the equalizer to "0." At this position you are neither boosting nor cutting the volume of any frequency. It is not necessary to reset the frequency controls because they don't do anything if the volume controls are set to "0."

Even if the EQ has an on/off switch, the volume knob should still be set to "0" so that when the EQ is turned on, it makes no changes and you are starting from "0," not some unknown preset.

3. Boost the Volume All the Way

Boost the volume on the band of EQ where you think the problem is. When first starting out, it is a good idea to boost the volume all the way. Be careful, though. Boosting the volume all the way in the bass area can blow up your speakers. And boosting the volume all the way in the midrange can make you deaf. It's a good idea to keep your other hand on the master volume control on your mixer or interface so as not to hurt yourself.

Visual 81.
EQ Knob with Volume
Set to "0"

Boosting the volume all the way will help you locate the frequency you want to turn up or down. A good analogy is when you cook with new spices. It's like putting your finger in the spice jar and testing it. By tasting it at full strength, *you know what that spice is about, big time*! Of course, you won't be boosting the volume all the way in the mix, but it makes it easier to find the frequency that you either want to cut or boost later.

If your plans are to cut a frequency, you can try cutting the volume all the way instead of boosting the volume all the way. Doing it this way is a little less annoying because you are looking for "good" sounds instead of irritating or muddy sounds. However, you run the risk of not finding the exact frequency that was the problem in the first place. It is only a good idea to use this technique when you already really know in your head what an instrument should sound like naturally.

4. Set the Bandwidth

The bandwidth (labeled "Q" on the knobs) is the range of frequencies that you are turning up or down.

The bandwidth is set to "thin" when cutting mudd, boosting lows, and cutting irritation in the midrange. It is set to "medium" when boosting highs.

When cutting mudd or boosting lows we set the bandwidth to quite thin so that it doesn't affect the neighboring frequencies. The muddy range is right next to the lows so if the bandwidth is too wide you will inadvertently affect your neighbor. For example, boosting the lows with too wide of a bandwidth will actually bring up the neighboring muddy range.

Visual 82.
Wide Bandwidth on
250 Hz Cut

The "thin" bandwidth I recommend is a little wider than the thinnest possible. On Pro Tools' "Dyn 3 EQ" this is six on the bandwidth setting. However, this number is not the same on other EQs. Here's what almost all the way thin looks like.

On most analog consoles, set the bandwidth as thin as it will go. On most digital consoles and digital audio workstations, all the way thin is *too* thin. Many digital EQs begin to "whistle" when the bandwidth is set too thin.

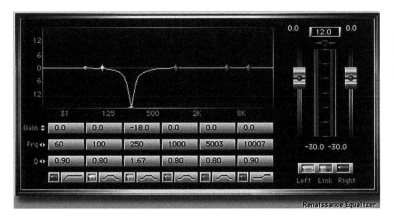

Visual 83.
Thin Bandwidth on
250 Hz Cut

Again, when boosting lows use a thin bandwidth so you don't boost the mudd back up.

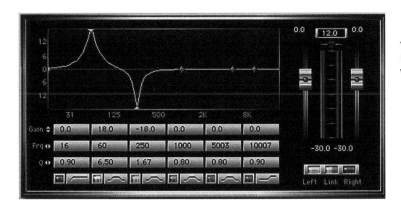

Visual 84.
Use a Thin Bandwidth
When Boosting Lows

When getting rid of irritating frequencies, set the bandwidth to almost all the way thin for a similar reason as above. If you were to use a wide bandwidth on a vocal, guitar, or cymbal, you will lose the entire body of the sound in the midrange – again because we are so sensitive to the midrange. Then the sound would be dull and not present.

When boosting highs, set the bandwidth to medium wide. This sounds more natural. This is three on the Bandwidth knob in Pro Tools' EQ. No telling what number on other units.

> After you have found and cut or boosted a frequency, you could try playing with the bandwidth to see if it sounds better a little wider or thinner.

5. Sweep the Frequency Setting

With the volume boosted all the way, sweep the frequency control to find the frequency you want to turn up or down. When your plans are to cut a frequency (as with mudd or irritation), then you are looking for the frequency that sounds the worst – the muddiest or most irritating. On the other hand, when trying to find a frequency to turn up (as with brightness or lows), you are looking for where it sounds the best.

I like to divide the sweep range into multiple frequency settings. For example, if I am working mudd in the 100–400 Hz range, then I will try 400, 350, 300, 250, 200, 150, and 100. Doing it this way makes it easier to compare which frequency is the worst (when cutting) or best (when ultimately boosting). Then after you decide which one of these to use, then sweep around that particular frequency to refine it even more. This technique for sweeping is even more helpful when EQ'ing something that only plays for very short durations such as toms, cymbals, or synth effects.

6. Return the Volume Knob to "0"

With the volume boosted all the way while listening to an annoying frequency that you are wanting to take out is very intense. You are listening to the worst sound ever turned up all the way. Commonly, a person loses touch with the reality of what the natural sound was in the first place. To regain your perspective on the tone of the sound before it was EQ'd, return the volume knob to "0" (on the EQ band you are working on). *Don't touch the frequency control.* You have just spent time boosting the volume all the way and sweeping to find the frequency. Don't lose your setting. Only return the volume control to "0." I find it easiest to click on the numbers in the Gain setting and type in "0." Or on some systems you can click on the knob while holding down Option (or Alt) it will go back to "0."

Also, don't just turn your EQ off – actually return the volume to "0."

Now listen to the sound and regain your grip on reality that you started with.

7. Boost or Cut the Volume to Taste

If you are turning a frequency down, slowly turn the volume down as much as you think it needs. If you are boosting a frequency, same thing. Play with the volume knob until you figure out how much it needs to be cut or boosted.

As with the sweep knob, I like to try four different volume settings instead of sweeping the volume knob. For example, when cutting the volume, I will cut 1/4 of the way, 1/2 of the way, and 3/4 of the way, then all the way and compare each as I go. Again, this is especially helpful when EQ'ing something that only lasts for a second.

8. Check to See if You Like What You Did

Turn the Bypass switch on and off, compare the EQ'd sound with the original sound, and make sure you like what you did (or if you can hear any difference at all).

When you EQ a second frequency range, only bypass that one frequency range instead of the whole EQ Bypass. That one you can see if you like just that one EQ that you have just done.

In the end, after having EQ'd multiple frequency ranges, hit Bypass to bypass all of the EQs on that one sound to see if you like what you did.

So far I have provided you with an extensive overview of how to use EQ. However, it requires practical experience to get know its intricacies. For those of you who are just beginning, here is a listing of common EQ techniques for well-known instruments – although, in reality, every sound is different.

Frequency	40–100	100–400	400–1000	1000–5000	5000–8000	8000–20,000
Sounds						
Bass	Bottom	Roundness, Muddiness	Body on Small Speakers	Highs, Presence		Hiss
Kick	Bottom	Roundness, Muddiness		Thud	Click	Hiss
Snare	X	Fullness, Muddiness			Brightness, Clarity	
Toms	X	Fullness, Muddiness		Presence	Brightness, Clarity	
Floor Toms	Bottom	Fullness, Muddiness		Presence	Brightness, Clarity	
Hi-Hat, Cymbals	Bleed	Fullness, Muddiness, Bleed		Irritation	Clarity, Crispness	Shimmer, Sizzle
Voice	Rumble	Fullness, Muddiness	Honkiness	Presence, Irritation, Telephone	Clarity, Crispness, Sibilance-6K	Sparkle
Piano	Bottom	Fullness, Muddiness	Honkiness	Presence	Brightness, Clarity	Harmonics
Harp	Bottom	Muddiness, Pedal Noise,			Brightness, Clarity	
Electric Guitar	X	Fullness, Muddiness, Crunch		Twanginess, Cut/Shred, Irritation	Thinness	Hiss
Acoustic Guitar	X	Fullness, Muddiness			Brightness, Clarity	Sparkle
Organ	Bottom	Fullness, Muddiness		Presence	Brightness	
Strings	X	Fullness, Muddiness		Irritation, Scratchiness	Clarity, Crispness	Shimmer
Horns	X	Fullness, Muddiness	Roundness		Clarity, Crispness	
Conga	Boominess	Fullness		Presence	Clarity, Crispness	
Harmonica	X	Fullness		Presence		

Chart 4. Equalization Chart

Frequency	Hi-Hat	Kick	Snare	Overheads	Toms	
High Highs (8–20K)	+3			+3		
Highs (5–8K)		+6	+6			
Midrange (1–5K)		+6			+6	
Low Mids (100–400)	–12	–10		–12	–6	
Bass (40–100)	–12		+2	–12		
Low Bass (<40)	–12	+2		–12		
	Bass	Distorted Guitar	Clean Guitar	Acoustic Guitar	Piano	Vocals
High Highs (8–20K)				+3		
Highs (5–8K)			+3	+3	+3	+2
Midrange (1–5K)	+5	+3				
Low Mids (100–400)	–3			–5	–3	
Bass (40–100)						
Low Bass (<40)	+2					

Chart 5. Common Quick General EQ

Hi-Hat and Cymbals		
Lows	Mids	Highs
Roll off whole low end up to 300 Hz	If irritating, find & roll off irritating frequency	Boost 12K 3 dB for sizzle. Watch out for irritation
Kick Drum		
Lows	Mids	Highs
Roll off mud between 100 and 300		Boost around 1–3K for thud, around 3–6K for click
Snare Drum		
Lows	Mids	Highs
Add 60–150 if it needs bass	Take out irritation or add edge depending on style of music and song	Add 3–10 dB around 3–6K
Toms		
Lows	Mids	Highs
Cut mud around 100–300		Add 3–8 dB around 3–6K
Bass Guitar		
Lows	Mids	Highs
Cut mud around 100–300 if needed, Add lows around 40–60	Boost around 1–5K for presence. Watch out for string noise.	
Electric Guitar		
Lows	Mids	Highs
Cut mud around 100–300	Add 1–3 for edginess. Watch for irritation.	
Acoustic Guitar		
Lows	Mids	Highs
Cut mud around 100–300		Boost 5–8K for Sparkle
Piano		
Lows	Mids	Highs
Cut mud around 100–300	Cut any honkiness around 600–1K	Boost 5–8K for Sparkle
Vocals		
Lows	Mids	Highs
Cut or boost 100–300 depending on mic, voice and mix	Listen closely for un-naturalness or megaphone and cut if so	Boost 3–6K for Presence and Clarity
Horns		
Lows	Mids	Highs
	Beware of irritating or honky frequencies. Cut if needed.	

Chart 6. Typical EQ for Typical Instruments

Common Terminology for EQ Frequencies
Even if you learn all of the frequencies and master how to EQ an instrument for different types of music and songs, the people you are working with might still be using street terminology to describe what they want. Therefore, Chart 7 is a list of slang and what it means.

<40	40–100	100–400	400–1000	1000–5000	5000–8000	8000–20,000
Bottom Ballsy The Deal Powerful Thumpin' Solid Beefy Fullness	Round Warm Fat	Warm Body Crunchy	Natural	Presence Projected Forward Intelligible Articulate Clear	Presence Clean Airy Bright Brilliant Live Smooth Crispy	Presence Crispiness Sparkle Sharp
Too Much Heavy Rumbly	Too Much Muddy Tubby Chunky Woofy	Too Much Muddy Barrelly Woofy	Too Much Megaphone Boxy Woody	Too Much Megaphone Hornlike Phonelike Honky Nasaly Edgy Steely	Too Much Metallic Strident Cutting Piercing Shrill Screamin'	Too Much Sizzling Searing Glaring Glassy
Not Enough Thin Wimpy	Not Enough Thin Anemic	Not Enough Distant Hollow Disembodied Thin Tinny	Not Enough Unnatural	Not Enough Veiled Covered Muffled	Not Enough Dull Dead Dark	Not Enough Flat Cheap

Chart 7. Common Terminology and Slang

Section C: Panpots and Stereo Placement

When mixing, you use panpots (balance knobs) to place each sound and effect left to right between the speakers. A panpot is actually two volume controls in one. When you pan to the left, the signal going to the right is turned down. When you pan to the right, the volume of the signal going to the left is turned down.

As previously discussed, panning in a mix is mapped out visually as a function of left to right. Panning, a sound to one side or the other also seems to make the instrument just a little bit more distant in the mix. If the sound is panned to the center, it will seem to be a little bit closer, a little more out front.

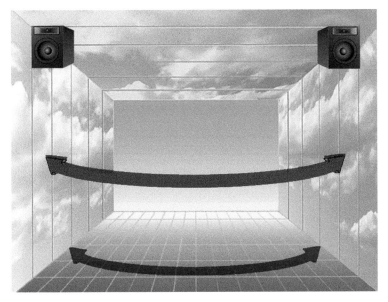

Visual 85.
Left and Right
Placement

If we think of the space between the speakers as a pallet on which to place instruments left to right, the main objective might be to place each sound in a different place so you can hear each sound more clearly. However, certain styles of music have developed their own traditions for the particular placement of each instrument left to right in the stereo field. Normally, the placement of a sound is static; it stays in the same place throughout the mix. However, the movement of a panpot during a mix creates an especially dramatic dynamic. We will discuss the common ways panning is used to create musical dynamics in the next chapter.

Section D: Time-Based Effects

Delays

After many failed attempts to use outdoor racquetball courts to create delays, engineers realized they could get a delay from a tape player. You could hear a delay by recording a signal on the record head, then listening to the playback head 2 inches later. The delay time could be set by changing the tape speed. Engineers used this technique for years. There was also a popular unit called the Echoplex, which fed a piece of tape through a maze of tape heads at different distances, each giving different delay times. Not bad, but the problem with tape is that every time you record over it, you get more tape hiss.

Then came digital delays, which record the signal digitally onto a chip, then use a clock to tell the unit when to play the sound back.

Delay Times Versus Distance

Before we explore different delay settings, it is helpful to understand the relationship between delay time and distance. Sound travels at approximately 1,130 feet per second. That's around 770 miles per hour, which is extremely slow compared to the speed of sound in wires – 186,000 miles per second, the speed of light (approximately 670 million miles per hour). Therefore, it is easy to hear a delay between the time a sound occurs and the time it takes for a sound to travel to a distant wall and back. We can also easily hear a delay when we put two microphones at two different distances from one sound. In fact, changing the distance between two microphones is almost exactly like changing the delay time on a digital delay.

> There are no delays in outer space. There is no sound in space. No air to carry the sound!

The following chart illustrates how different distances relate to delay time. Of course, if you are calculating a delay time based on the distance between a sound source and a wall, the distance must be doubled (to and from the wall).

As distances become smaller and smaller, the distance in feet almost equals the milliseconds of delay. This correlation comes into play when using more than one mic on a sound (e.g., piano, guitar amps, acoustic guitars, horns, or background vocals) and is especially helpful when miking drums. For example, the distance you place overhead mics above the drum set will create a corresponding delay time between the overhead mics and the snare mic (or any of the rest of the mics for that matter). It is also important to note the distance between instruments when miking an entire band live (or recording everyone in the same room at once) since mics may be more than 10 feet away from another instrument and still pick it up.

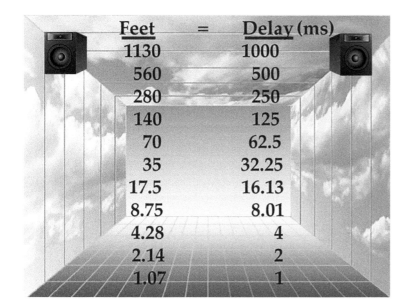

Chart 8.
Distance Versus
Delay Time

Feet	=	Delay (ms)
1130		1000
560		500
280		250
140		125
70		62.5
35		32.25
17.5		16.13
8.75		8.01
4.28		4
2.14		2
1.07		1

Besides delay time, you must also consider phase cancellation, a problem that happens with extremely short delay times. We'll discuss more about this later.

If you pay attention to the way that something sounds when miked at different distances, you will eventually become aware of what different delay times sound like. Once you become familiar with the way that different delays affect different sounds, you can control their use in a way you deem most appropriate; that is, you can do whatever you want.

Different Delay Times

> You need to learn how
> each delay time feels and
> what feelings or emotions each delay time evokes.
> What you are doing when you use a delay
> is applying a certain feeling to the song.

> There might be very few rules in this industry, but this is a good one: gain a perspective so that you know what you are doing. Then, if anyone disagrees, it doesn't matter.

Let's define specific delay time ranges so that you can get to know them and incorporate them into your memory time banks.

Echo – More Than 100 ms

Professional engineers refer to this length of delay as *echo*. However, the real world (and my mom) use the term echo to refer to reverb. For our purposes, we will use echo to refer to a delay time greater than 100 ms, not reverb.

When setting a delay time greater than 100 ms, it is important that the delay time fits the tempo of the song, otherwise it will throw off the timing of the song. The delay time should be in time, a multiple of, or an exact fraction of the tempo. If you know the beats per minute of the song, the following chart gives the relationship between tempos and delay times. However, most delays have a setting that Synchs the delay time to the tempo of the song, which is much easier.

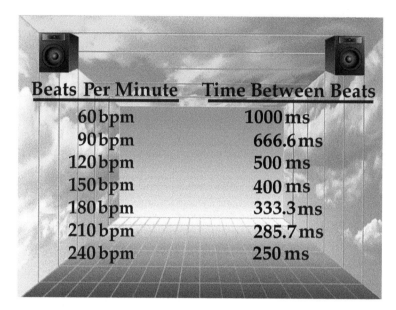

Beats Per Minute	Time Between Beats
60 bpm	1000 ms
90 bpm	666.6 ms
120 bpm	500 ms
150 bpm	400 ms
180 bpm	333.3 ms
210 bpm	285.7 ms
240 bpm	250 ms

Chart 9.
Tempo Versus
Delay Time

If you know the tempo of the song, you can figure out the delay time with the following formula:

Delay time = 60,000/beats per minute

Then, any fraction or multiple of that delay time will also fit the tempo of the song. For example, if the tempo is 100 BPM then 600 ms would fit the tempo. But 150 ms, 300 ms, and 1,200 ms would also fit.

If you don't know the beats per minute (bpm) of the song, use the snare drum (or some other instrument playing a continuous pattern) to set the delay time. Even if you are going to put the delay on the vocals, for example, put the delay on the snare to set the delay time to the tempo. It is also really helpful to pan the snare separately than the delay so you can tell which is which. Again, once you have found a delay time that works, any multiple or fraction of that time might also work.

A delay time over 100 ms creates a dreamy effect and is most commonly placed in songs with slower tempos where there is room for the additional sound. Therefore, the more instruments and the more notes in a mix, the less often this delay time is used. This is pretty obvious – if you have no room in the mix, don't add more sounds. This is especially true when there is feedback on a long delay time. The delays take up so much space in a mix that they are often only turned up at the end of a line, where there is enough space to hear the echoes by themselves.

Feedback is created by feeding back the delayed signal into the input, so the sound repeats, repeats, repeats.

Slap – 60–100 ms

You can hear this delay time, commonly referred to as *slap*, on the vocals of Elvis Presley and in rockabilly music. In fact, there is about an 80 ms delay between the syllables "rock" and "a" in the word "rockabilly."

This effect can be quite helpful in making a thin or irritating sound (especially a voice) sound fuller. It can help to obscure bad vocal technique or pitch problems. In fact, a slap can be used to bury any bad sound. However, you never want to bury anything too deep. Add too much delay on a bad vocal and not only do you have a bad vocal, but you also have a bad mix.

A slap can make a vocal seem less personal but it can also just be a cool effect for some songs, depending on the style of music. It is also really nice to put a slap on a group of vocals as it makes it sound like there are twice as many.

If I have an incredible singer who is transmitting some really high energy, I normally refrain from using a slap so that I can hear the beauty of the nuances in their voice and performance.

Doubling – 30–60 ms

Put your lips together and blow a raspberry (this is the interactive portion of the book), technically called a "motorboat." The time between each flap of your lips is approximately 50 ms. Delay time in this range is referred to as "doubling" because it makes a sound seem like it was played twice, or double tracked. When a part is sung or played twice there will naturally be a time delay ranging from 30 to 60 ms (it really hard to sing or play a part twice exactly in time). Therefore, adding a delay of this length makes it sound like the part has been played twice. The Beatles used this effect extensively to simulate more vocals and instruments.

Doubling is also good on a large vocal section. It can easily double the size of a choir. I like to use both doubling and slap to create that Mormon Tabernacle choir effect. In fact, I might use four delays: 30, 45, 60, and 75 ms.

Just like a slap, doubling helps to obscure a bad sound or a bad performance. So it can be used to help blur a not so great sounding voice or performance. Likewise, since it does obscure the purity and clarity of a sound, you should use it selectively, depending on the sound, song, and style of music.

> Although doubling makes a sound seem like it has been played twice, it is a different sound than if you actually doubletrack a sound. In fact, doubling often sounds so precise that it sounds somewhat electronic. This is especially true on vocals and simple sounds. However, if a sound is complex, especially if the sound is a combination of sounds (like a bunch of background vocals or a guitar sound with multiple mics), then you don't notice the precision of the delay. Therefore, when you put doubling on twenty vocals, it sounds like forty vocals, and it sounds incredibly natural.

Fattening – 1–30 ms

An unusual thing happens with this type of delay, commonly known as fattening. At this delay time, our brain and ears are not quick enough to hear two sounds; we only hear one fatter sound.

The threshold between hearing one sound versus two sounds actually varies depending on the duration of the sound being delayed. Also, the further the sounds are panned separately, left and right, the *shorter* the delay time before you here two sounds. For example, a guitar panned to the center with a delay in the center might require at least 40 ms to hear two sounds; whereas, if the guitar and delay are panned left and right, you might hear two sounds beginning around 20 ms. The following chart gives approximate thresholds for some instruments with different durations (actual thresholds will depend on the particular timbre and playing style of the instrument):

Chart 10.
Quickness of Brains

APPROXIMATE THRESHOLDS BETWEEN HEARING ONE SOUND VERSUS TWO

Hi-hat	10 ms
Percussion	10 ms
Snare	15 ms
Kick Drums	15 ms
Piano	20 ms
Horns	20 ms
Vocals	30 ms
Guitars	30 ms
Bass Guitars	40 ms
Tubas	40 ms
Slow Strings	80 ms

Besides reverb, fattening is the most-used effect in the studio, mostly because it doesn't sound much like an effect. Fattening is the primary effect used to make a sound stereo, which has a certain magic to it. When you put the original "dry" instrument sound in one speaker and put a delay less than 30 ms in the other speaker, it "stretches" the sound in stereo between the speakers.

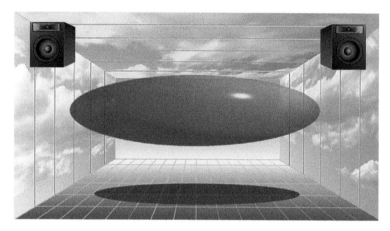

Visual 86.
Fattening: Delay
<30 ms

Fattening is very effective in making a thin or irritating sound fatter and fuller. It also makes a sound more present simply because when a sound is in stereo, it takes up more space between the speakers. This is especially effective when you want to turn a sound down in the mix but still have it be discernible.

You have to be careful with fattening, though, because it uses up your space between the speakers. Fattening will make a mix fuller and denser, so you must make sure there is enough room between the speakers. Therefore, fattening is used most often when there are fewer notes and sounds in the mix. On the other hand, if you want to create a wall of sound, even if the mix is already busy, you can add fattening to make it more busy. This is commonly done in heavy metal, alternative rock, and even some new age music.

> Because fattening makes a sound bigger and wider between the speakers
>
> (when panned) it makes a sound "more" of whatever it is already.
>
> If a sound is beautiful, it makes it more beautiful.
>
> If a sound is intense, it makes it more intense.
>
> If sound is annoying or has a buzz in it, it is now really annoying.

I prefer to use two delays for fattening. And, I have found that 12 and 18 ms work for most sounds. These are a good place to start with when using fattening.

Phase Cancellation – 0–1 ms

Putting 0–1 ms on a sound changes the tonality of the sound just a bit, so can be interesting to play with. It can also be used to fix phase cancellation on an isolated sound.

Phase cancellation can get quite complex so I'll only address the critical aspects here. But keep in mind that phase cancellation is a very serious problem in recording and I highly recommend that you do further research to gain a complete and clear explanation of the problems it causes.

Phase cancellation happens when two of the exact same sound, like those created with two mics or two speakers, are a little bit out of time. One example is when you switch the positive and negative wires on one of two speakers. Now, one speaker is pushing out while the other is pulling in. When a speaker pushes out, it creates denser air than normal. When a speaker pulls in, it creates more spaced out air than normal (rarefied air). When the denser air from one speaker meets the spaced out air from the other speaker, you end up with normal air; normal air equals silence. This means you could have two speakers blasting away and theoretically you could hear nothing.

There are many companies now using phase cancellation to quiet the world. This technology is used in automobiles, on freeways (instead of cement walls on the sides of the freeways), in factories, and even in headphones to cancel out sounds around you. Marriage counselors are selling them by the dozens.

If you have two mics on one sound at two different distances, one mic might be picking up denser air while the other mic is picking up spaced out air. Put the two mics together in the mix and they will tend to cancel each other out, though not completely. Phase cancellation degrades the sound quality in the following ways:

1. *Loss of volume.* You lose volume when both mics are on, especially when you're in mono (which, by the way, is one of the best ways to detect phase cancellation – put the board in mono or pan both sounds to the center).
2. *Loss of bass.* You lose bass frequencies, making the sounds thin.
3. *Loss of image.* Most importantly, you lose the clarity and precision of the perceived image of the sound between the speakers. The sound seems to be more "spacey." Though some people like this effect, most people are addicted to clarity these days. If the mix is ever played back in mono (as on TV or AM radio), the sound will disappear completely.

There are many ways to curb phase cancellation. The primary way is to simply move one of the mics. If both mics are picking up the sound in the same excursion of the wave, there will be no phase cancellation.

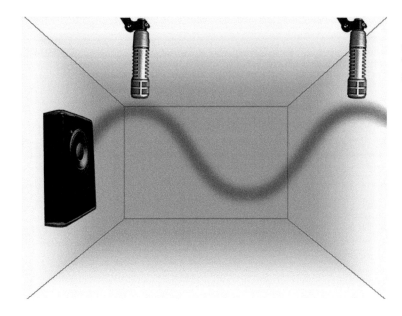

Visual 87.
Two Mics Picking Up
Sound in Phase

It takes 1 ms for a complete wave of 1,000 Hz to pass by us. If you were to set a delay time of 0.5 ms on a sound, it would put it out of phase. Therefore, you can use a digital delay of less than 1 ms to put the sound back in time.

Finally, you can remove a large amount of phase cancellation through isolation. Often, the bleed of a sound into a second mic will cause phase cancellation with the first mic. By using baffles or noise gates, you can reduce the bleed in the second mic, voiding the phase cancellation.

Panning of Delays
When the delay time is long enough to hear two sounds, then the delayed signal can be treated just like another sound and can be placed anywhere in the mix using volume (front to back), panning (left to right), and EQ (up and down).

Visual 88.
Volume, Panning, EQ,
Movement of Delay
>30 ms

When the delay time is less than 30 ms or so, fattening occurs. You can also place this line of sound anywhere with volume, panning, and EQ.

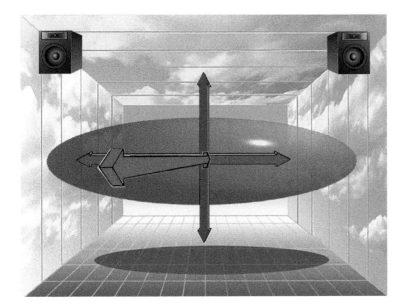

Visual 89.
Volume, Panning, EQ, Movement of Fattening

Flangers, Choruses, and Phase Shifters

In 1957, Toni Fisher was doing an album and someone accidentally brushed up against one of the reels on the tape player, slowing down the tape. When the person stood up, it sped back up to normal speed. The band said, "Cool, let's put it on the record." They did put it on the record, and thus flanging was born. The song, "The Big Hurt," went to number 3 on the charts in 1957. For years, engineers created flanging by putting their fingers on the metal "flanges" that hold the tape reel together to slow the tape down, thus the term *flanging*. Many blisters later, we realized that you could create the same effect using a digital delay.

If you set a digital delay for less than 30 ms of delay time and crank up the feedback, you get an effect called tubing (check it out on a digital delay). The interesting thing is that the shorter you set the delay time, the higher the pitch of the tube. The longer the delay time, the lower the pitch of the tube. Now, if you set a clock to sweep the delay time back and forth between, say, 9 and 1 ms, then you get the effect called flanging.

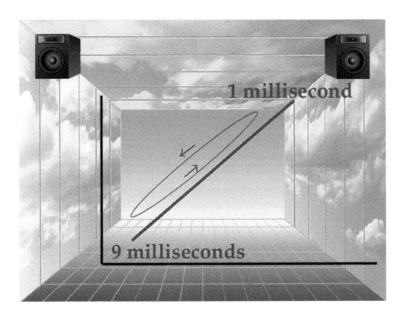

Visual 90.
Pitch Versus Delay Time of Flanging

Using the visual framework, flange is shown like this:

Visual 91.
Virtual Mixer
Flanging

If you set the range of the sweep (called width, depth, or intensity on different units) so that the sweep of the delay time is not so wide (say 2–3 ms), you then have the effect called chorusing. (Chorus effects have a delay like doubling or fattening also added.)

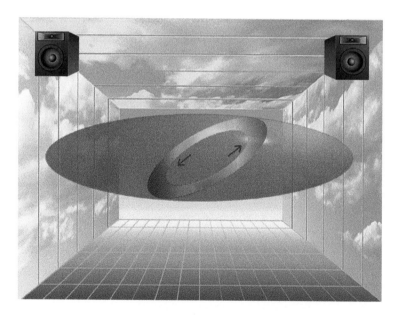

Visual 92.
Virtual Mixer
Chorusing

If you set the delay time so that you are only sweeping between 0 and 1 ms, you hear the effect called phasing.

There are various parameters or settings found on flange, chorusing, and phasing units:

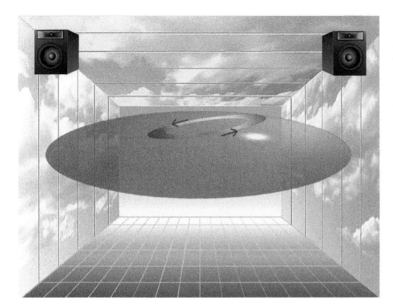

Visual 93.
Virtual Mixer
Phasing

Rate, Speed, or Frequency

The setting is the time it takes for the delay to sweep back and forth between two delay times. For example, it can be set to take 1 second to smoothly change from 1 to 9 ms and back. The rate of the sweep can be set to the tempo of the song – you might have it rise on one beat and fall on the next beat – or to rise on one chord and fall on the next chord. You could even set it to rise on the first half of the verse and fall on the second half. The rate is often set so slow that it doesn't correspond to any part in the music.

Width, Depth, or Intensity

This setting is the range of the delay sweep. For example, a narrow width setting might sweep between 2 ms and 3 ms, while a wide width setting might sweep between 1 ms and 9 ms. Because pitch corresponds to the delay time, this means that the wider (or deeper) the setting, the wider the frequency change.

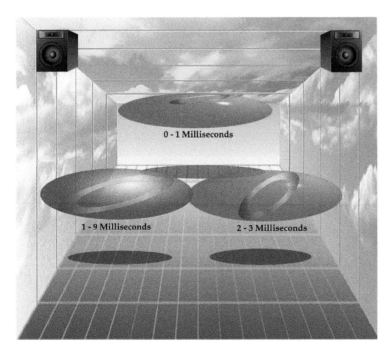

Visual 94.
Narrow and Wide
Sweep on Flange,
Chorus, or Phase Shifter

Feedback

Feedback takes the output of the delay and "feeds it back" into the input. Some feedback is required to get the flange effect in the first place. The more feedback you add, the more intense or dynamic the frequency sweep.

Negative Feedback

Negative feedback puts the signal being fed back into the input out of phase. This generally causes a more hollow tubular type of flange sound.

Each of these effects, flanging, chorusing, and phasing, has short delay times with feedback, with the delay time sweeping back and forth between shorter and longer delay times. They don't normally make things clearer and more pristine. They tend to obscure a sound, which can often be a totally desirable effect. Because they are unusual, they can make a sound stand out in the mix. Although, these days, they have been used so much that they are not that unusual anymore.

Flanging is used to create a more spacey type of mood, an other-worldly effect. It's great for making things sound like they are under water. Chorusing is often used to simulate a chorus of people or chorus of instruments. It simulates the effect of a group of people going ever so slightly in and out of pitch. The sweep in pitch is at a very high frequency on phasing, so it is a very subtle effect – so subtle that when used at Grateful Dead concerts, the crowd often wondered if the effect was actually coming from inside their heads. When phasing is panned left and right with a very short delay sweep (0–0.1 ms) you can get a very cool effect where the sound sweeps in a 3D circle between the speakers. In headphones, the sound sweeps in a circle around your head.

All of these effects are not grounding!

Each of these effects can be panned in various ways:

Visual 95.
Flanging Panned
Various Ways

Each can also be brought out front with volume . . .

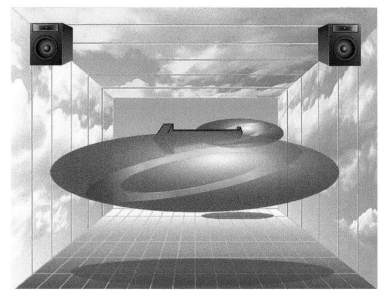

Visual 96.
Flanging at Different
Volumes

. . . and raised or lowered a little bit with EQ.

Visual 97.
Flanging EQ'd
Differently

Reverb

Reverb is hundreds and hundreds of delays. When a sound first occurs, it travels throughout the room at the snail's pace of around 770 miles per hour. It bounces off the walls, ceiling, and floor and comes back to us as hundreds of different delay times. All of these delay times wash together to make the sound we know as reverb.

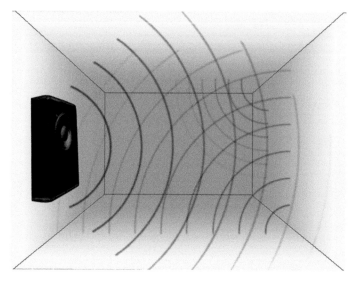

Visual 98.
Waves Bouncing Around
Room

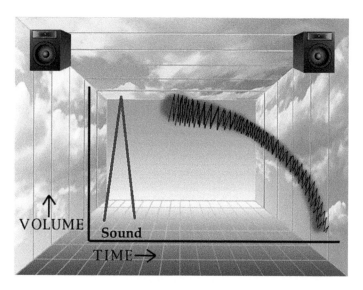

Visual 99.
Wash of Reverb

When you place reverb in a mix, it is like you are placing the sound of a room between the speakers. Therefore, reverb is shown as a room or cube between the speakers.

Visual 100.
Virtual Mixer Reverb

Remember though – reverb is actually like placing hundreds of spheres of sound between the speakers. It takes up a tremendous amount of room in this limited space between the speakers. In a digital reverb, all of these delays are panned to virtually hundreds of different places between the speakers. This is why reverb masks other sounds so much in the mix.

Visual 101.
Reverb: Hundreds of Delays
Panned Between Speakers

There are certain parameters of control found in units that create reverb. I will explain each setting and show it using the visual framework.

Room Types

Modern digital reverbs allow the user to change the "type of room." You can simply imagine different types of rooms between the speakers. There are no strict rules as to the type of room that is used in a mix. Some engineers prefer a plate reverb sound on the snare drum. Some use hall reverbs on saxophones.

It is best to always set the type of reverb while in the mix (with all the sounds on) to make sure it cuts through the mix like you want it to. Different types of sounds will mask the reverb in different ways.

Reverb Time

You can also change reverb time: the duration or length of time it lasts. On some units it is called "decay."

Visual 102.
Long and Short Reverb
Times

A common rule is to set the reverb time on a snare drum so that it ends before the next kick lick; this way, the snare reverb does not obscure the attack of the next kick note, which will keep the kick drum sounding clean, punchy, and tight. The faster the tempo of a piece, the shorter the reverb time. Again though, rules are made to be broken. (You won't go to jail for this one.)

Predelay Time

When a sound occurs, it takes a while for the sound to reach the walls and come back. The time of silence before the reverb begins is called the *predelay time*. On many units it is just called *delay*.

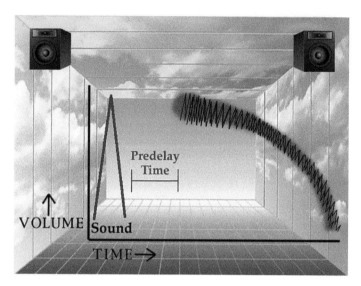

Visual 103.
Predelay Time

Different sized rooms will naturally have different predelay times. A medium-sized auditorium has around 30 ms of predelay time, while a coliseum might have as much as 100 ms of predelay time. Therefore, it is important to have a bit of predelay time if you are looking for a truly natural reverb sound. Most times, when you call up a preset in a reverb unit, someone has already programmed a predelay time. You can adjust this as desired.

The cool thing about longer predelay times (over 60 ms or so) is that they help to separate the reverb from the dry sound. With shorter predelay times, reverb will very quickly "mush up" the original dry sound, making it unclear. With longer predelay times, a vocal, for example, will remain clean and clear even with a good amount of reverb. When using extremely long predelay times, it is important to set the delay time to the tempo of the song (as was covered when I discussed delays).

Diffusion

In most effect units, diffusion is the density of the echoes that makes up the reverb. Low-diffusion has fewer echoes.

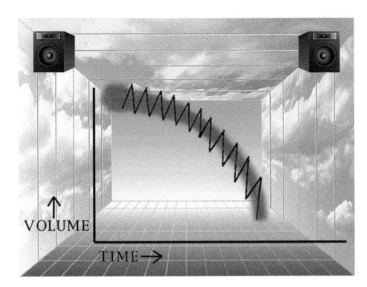

Visual 104.
Low-Diffusion Reverb

You can actually hear the individual echoes in a low-diffusion setting. It sounds kind of like "wil, il, il, il, il, bur, bur, bur, bur, bur, bur." A hall reverb setting is preset with a very low-diffusion setting. High-diffusion has more echoes – so many that they meld together into an extremely smooth wash of reverb. Plate reverbs often have a very high-diffusion preset.

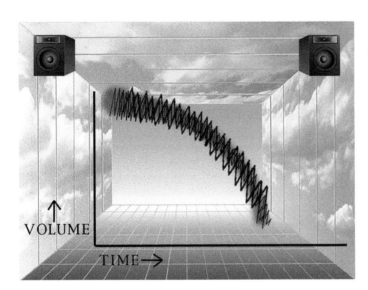

Visual 105.
High-Diffusion Reverb

There are no strict rules for the use of high- or low-diffusion settings. High-diffusion settings tend to be sweeter, smoother, and more silky. Low-diffusion tends to be more intense. Some engineers prefer a low-diffusion setting on a snare drum to make it sound more raucous for rock and roll. High-diffusion is often used to make vocals sound smoother.

EQ of Reverb

You can equalize reverb at various points in the signal path. First, you can EQ the reverb after the signal comes back into the board (if you are using channels for your reverb returns that have EQ on them). It is usually better to use the EQ in the reverb unit itself. Not because it is necessarily a better EQ, but because in some units you can place the EQ before or after the reverb. Ideally, it is best to EQ the signal going to the reverb. If your reverb unit does not have this capability, you can actually patch in an EQ, after the master auxiliary send, on the way to the reverb unit. The truth is, I don't like to EQ reverb because it screws up the natural sound of a live space. If you couldn't care less about natural sounds (which is just fine) go ahead and use EQ on your reverb. Sometimes EQ might be used to simply roll off some low-frequency rumble. Normally, if your reverb sounds like it needs EQ, it is often better to go back and EQ the original sound that is going to the reverb.

High- and Low-Frequency Reverb Time

Even better than using EQ on your reverb is to set the duration of the highs and lows. Many reverb units have this setting these days. This is a bit different than EQ, which changes the volume of the frequencies. High- and low-frequency reverb time changes the time that each frequency range lasts. Using these settings will generally make the reverb sound more natural than any type of EQ.

Regardless of whether you EQ your reverb or set the duration, there is a huge difference as to how much space it takes up in the mix – and the resulting masking it creates. Remember that low-frequency sounds take up way more space than high-frequency sounds. And because reverb is also hundreds of sounds, reverb with more low frequencies will take up an enormous amount of space in a mix.

Visual 106.
Reverb with Low-Frequency EQ Boost

Reverb with more high frequencies still takes up a lot of space, but not nearly as much as when lows are present.

Visual 107.
Reverb with High-Frequency EQ Boost

Reverb Envelope

Another setting of reverb is the "envelope"; that is, how the reverb changes its volume over time. Normal reverb has an envelope where the volume fades out smoothly over time.

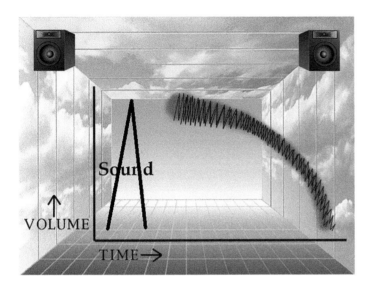

Visual 108.
Envelope (Change in Volume Over Time) of Normal Reverb

Engineers (being the bored people they are) thought to put a noise gate on this natural reverb, which chops it off before the volume has a chance to fade out. Therefore, volume stays even, then stops abruptly.

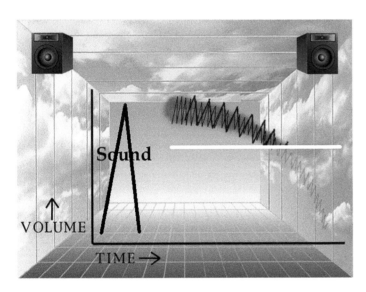

Visual 109.
Envelope of Gated Reverb

You can put a noise gate on your reverb, but it's much simpler to use the gated reverb setting on your effects unit. If we were to reverse the envelope of normal reverb, the volume would rise then stop abruptly.

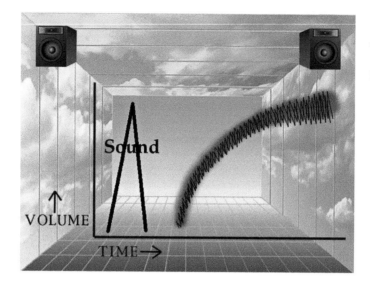

Visual 110.
Envelope of Reverse Gate
Reverb

Preverb is an effect that you can't buy because it puts the effect before the sound. Originally, it was created playing the tape backwards and recording reverb on one of the tracks, then turning it back around forward. You can create the effect digitally, by finding the function to reverse a sound (in AudioSuite in Pro Tools). Then record reverb on the backwards sound, and reverse the whole thing again.

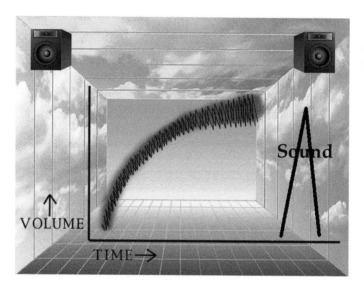

Visual 111.
Preverb

This effect is the most evil one that can be created in the studio; only the devil could put an effect on something before it happens. Furthermore, it has been used in every scary movie made, including *The Exorcist* and *Poltergeist*. And, of course, it is one of Ozzy Osbourne's favorite effects.

One of reverb's main functions is to connect sounds in a mix and fill in the space between the speakers.

Visual 112.
Reverb Filling In Space
Between Speakers

Like any sound, reverb can be panned in various ways.

Visual 113.
Reverb Panned to Left

Visual 114.
Reverb Panned from
Left to 1:00

Reverb can be spread to any width by how far left and right you pan the reverb return channels on your mixing board. It is best to use an Auxiliary Send going to an Auxiliary Track with reverb on it so you can pan the reverb separately than the original dry instrument.

Visual 115.
Reverb Panned from
11:00 to 1:00

Visual 116.
Reverb Panned from
10:00 to 2:00

Reverb can also be brought out front with volume . . .

Visual 117.
Reverb Turned Up in Mix

. . . placed in the background by turning down the volume . . .

Visual 118.
Reverb Turned Down
in Mix

. . . or raised or lowered a bit with EQ.

Visual 119.
Reverb with High-Frequency
EQ Boost

Visual 120.
Reverb with Low-Frequency
EQ Boost

Harmony Processors, Pitch Transposers, and Octavers

A harmony processor (harmonizer, pitch transposer, and octaver) raises or lowers the pitch, while keeping it in time. A harmonizer takes a longer, lowered pitch, deletes tiny slivers of sound (individual samples), and then splices it back together to keep it in time. This means you can have Darth Vader singing a happy song in time. A harmonizer also takes a shorter sound that has been raised in pitch, makes copies of the sound, and then splices them back together, putting it back in real time. Therefore, you can have the Chipmunks singing the blues in time with the rest of the band. Often, on cheaper harmony processors, you can even hear the "glitches" where the sounds have been spliced back together to put them in time.

When you raise or lower the pitch of a sound, it directly affects the amount of space it takes up. The higher the pitch, the less space the sound takes up.

Each and every effect has its own world of feelings that it brings to a mix. The trick is to get to know the feeling it gives you.

Pitch Correctors

A pitch corrector is similar to a harmony processor, but it can be used to fix the pitch in real time. They normally are not used to add harmonies because they don't change the pitch that much. They have revolutionized the industry making it so that many people can sing in tune who weren't able to before. The software version allows you to see the waveform and choose which words you actually want to use it on. For the most part, this won't affect the mix, other than making it sound like you have a better singer.

Section E: Other Effects

There are thousands of other effects out there. I don't claim to be a master at the whole perspective.

I should mention some interesting ones though.

Analog Effects

There is a large number of effects that are designed to simulate the analog effect of tape that most people agree is warmer. You might do a search for "analog effects."

There are also a huge array of "outboard effects" that are physical rack units. Generally an outboard unit is always better quality than a plugin because the entire unit is dedicated to one effect, whereas computers do multiple effects.

At the risk of upsetting a lot of people and manufacturers, here is a list of some highly recommended plugins from one of the top engineers on the planet. Of course, this list will be outdated pretty soon.

Mastering
"Studio 1" by Presonus for Mastering.

MacDSP Limiter – "6034 Ultimate Multi-band Limiter."

Effects
"VerbSuite" and other Plugins by Slate Digital.

"Seventh Heaven" and "Reverberate" by Liquid Sonics.

Processing

"Teletronix LA2A Leveler Collection," "Neve 88RS Channel Strip Collection," and "API Vision Channel Strip" by Universal Audio.

EQ by Sonible.

"FilterFreak" analog EQ by Sound Toys.

Backward Recording

Originally, we would record instruments while the tape is playing backward, or you can simply reverse the waveform in the computer to play backward. One of the coolest effects I remember recording was the strumming of a piano with a guitar pick – backward! We used it to start the song. I've recorded a whole range of instruments backward, including vocals (no evil messages though), guitar leads, piano, hi-hat, and shaker.

Section F: Combinations of Effects

Besides creating dynamics with each of the individual tools, you can also stack effects. Of course, you can add multiple effects to a mix, but *stacking* effects means to send the output of one effect to another effect. You can end up with some very unusual combinations of effects. Let's look at the following:

- Delay into reverb
- EQ into effects
- Chorus, flanging, and phasing into reverb

Delay Into Reverb

Send a sound out for delay, have it return on an Auxiliary Track. Then send the delay on the Auxiliary Track to the reverb. You can also try panning the delay separately from the reverb (both of which might be in a different place from the original dry sound). The volume of the delay compared to the reverb can be adjusted to make the nuances fit your taste. You could even EQ the delay differently from the reverb and from the original dry sound.

One of favorite effects (when appropriate) is to turn up the feedback on the delay before it is sent out to the reverb. However, instead of using a send, buss the output of the delay Auxiliary Track to the input of the reverb channel. What you end up with is reverb with a really long predelay and feedback. It is an extremely smooth effect that, when put on a vocal, makes it sound like you have a vocal synth in the background. Many guitarists, including David Gilmore of Pink Floyd, like the effect on their guitar as well.

EQ Into Effects

It is also nice to split a signal, send one to the mix, and then equalize the second signal going to the effect so the low end is rolled off. Send this high-end part of the sound to an effect such as reverb or flanging.

Chorus, Flanging, and Phasing Into Reverb

Another interesting effect is to send the output of flanging, chorusing, or phasing to a reverb unit. This is often much nicer than simply putting these effects and reverb together in the mix. Flanged reverb is especially effective when you have the flange rate set the tempo of the song, and you then pan the reverb around the room in time to the music (we're getting out there, now).

Another interesting effect is to slow a sound down or speed it up by using a time stretch function so that it is unintelligible and use it as a new sound. It gets really interesting when you then start sending that new bizarre sound through some unusual effects.

There are certainly a huge number of really cool combinations of effects. See if you can come up with some new ones on your own. If you come across anything that is death-defying – or just really nice – let me know (e-mail: David@GlobeRecording.com). Maybe I'll put it in the next edition of the book.

CHAPTER 5

Musical Dynamics Created With Studio Equipment

To make a great mix, you must determine what you can do in a mix. What dynamics can be created with the mix as opposed to the songwriting or production.

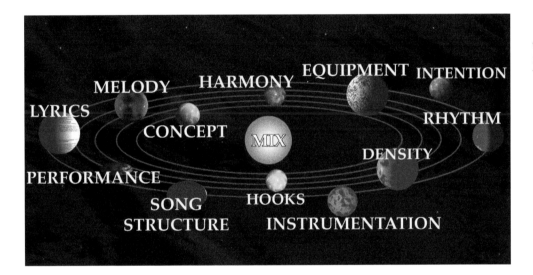

Visual 121.
Mix as Center of
Thirteen Aspects

When mixing, the four types of tools that you can use to create all the different styles of mixes in the world are volume faders, panpots, equalization, and effects.

> The art of mixing is
> the way in which
> the dynamics you create with the equipment in the studio
> interface with
> the dynamics apparent in music and songs.

When we speak about dynamics, we are not talking about the common terminology used for volume dynamics. We are not talking about changes in loudness. We are talking about changes in intensity. For example, this could be intensity of feelings, emotion, music, or structural. On a basic level, dynamics means anything that causes a change in us.

The Dynamics in Music and Songs

Before we explore the dynamics that can be created with studio equipment, let's explore the dynamics found in music and songs. A dynamic in music is anything that you get out of music. Music touches us in just about every aspect of our lives, and however you relate to music is, of course, valid. There are millions of dynamics discernible in music that affect us mentally, emotionally, physically, visually, psychologically, physiologically, and spiritually.

The most common dynamic that people feel in music is "up" and "down," whether it be on a physical, abstract, emotional, or psychic level. Some people feel very strong emotions when they hear certain types of music. It can make them happy or sad. It can crack them up with laughter or bring tears to their hearts.

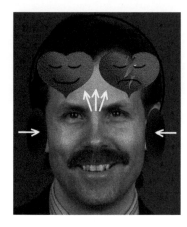

Visual 122.
Some People Get Feelings and Emotions out of Music

Some people see structure in music. And often they relate these structures to common structures found in the world, such as buildings, bridges, and pyramids.

Visual 123.
Some People See Structure and Form in Music

There are those who actually see the workings of the brain in a song. They see the way in which our minds work as being similar to the flow of a song. Other people even think of songs as thought forms. In fact, there are bands that write their music to represent the way the brain works. This explains the common theory that music is just an extension of our personalities.

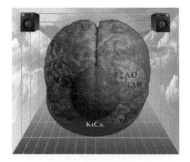

Visual 124.
Some People See Thought Forms and the Mind

Some people relate to music through music theory. They see notes on a scale, the intervals between notes, and chord structures. There are thousands of schools that teach the incredibly complex detail found in the study of music itself.

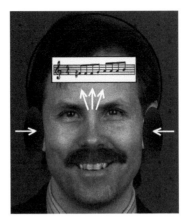

Visual 125.
Some People See Music Theory

Most of us also have physical reactions, like toe tapping, finger snapping, head bopping (or banging), and dancing. Much of the study of dance is how movement is related to music: shake, rattle, and roll. Physically, music can make us feel good from head to toe.

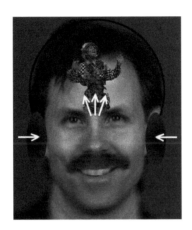

Visual 126.
Some People Move When They Hear Music

Not only does music move us physically, there is also a whole world of music therapy based on the healing vibrations of sound and music. Just imagine: if you could place instruments in a mix at different places in your body, where would you put the kick drum? How about the guitar or string section? Try a tuba in your tummy or sitar in your chest. Or how about reverb in your brain?

Visual 127.
Where in the Body Would You Put Sounds in Mix?

There is now a huge field of Sound Healing and Therapy that uses both sounds and music for a wide range of health issues including sleep enhancement, pain management, ADD/ADHD, PTSD, depression, anxiety, grief, and autism. Our institute is now the leader in this field – www.SoundHealingCenter.com. Most integrative therapy centers in major hospitals around the country are now incorporating Sound Healing and Therapy.

Some people see visual imagery or abstract colors and images. Walt Disney saw flying elephants.

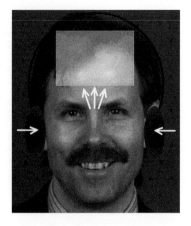

Visual 128.
Some Get Imagination Out of Music

Just check out MTV to see a whole other world of visual imagery. There are also those who see bubbles.

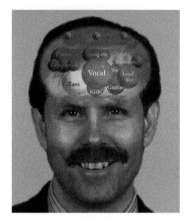

Visual 129.
Some See Bubbles

Then there are those who see spiritual connotations. The whole world of religious music is a good example. Music is often seen as a direct connection to God. Others go elsewhere.

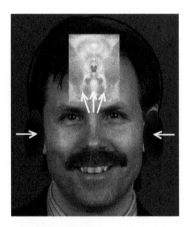

Visual 130.
Some See Spirituality in Music

So now you see that music can evoke a wide range of possible dynamics in people. They are as varied as people and life itself, and they are all valid.

> The job of an effective engineer is
>
> to create dynamics with the equipment
>
> that *fit*
>
> the dynamics in the music and songs. They can fit in various ways,
> but they should fit in some way.
>
> *But more importantly,*
>
> *the key is to use the equipment to enhance the magic.*

Magic in Music

It is really about the magic when you get down to it. What lights you up in the music you listen to? This is really the whole deal!

Magic could mean a large number of things: beauty, intensity, weirdness, emotional sincerity, unique, interesting, fun, or higher energies coming through. It is whatever turns you on.

Magic can manifest in any of the thirteen aspects. We'll discuss this in more detail in Chapter 7.

The Dynamics Created by the Equipment

So, what are the dynamics and magic that you can create with the equipment in the studio? Every mix in the world is created with the same four tools: volume faders, panning, equalization, and effects. Each of these four tools can be used to create some type of feeling that should fit or enhance the music and song, somehow. In order to explore the range of dynamics that can be created with each tool, I have identified three different levels of intensity, with Level 3 being the most intense.

Level 1 – Individual Settings

This is the difference between the individual levels or settings of each piece of equipment. Setting volumes, EQ, panning, and effects at specific levels creates a relatively minor emotional effect. For example, a lead vocal that is placed way out front from the rest of the mix by turning it up louder will be perceived completely differently than a vocal placed low in the mix. The message and the emotion will come across in a different way.

A lead guitar will be perceived quite differently if it is panned to the center as opposed to being placed far to the left or right.

A kick drum with a bright EQ that accentuates the click sound of the attack will be felt differently than the same sound with the highs EQ'd out and the lows boosted excessively.

A snare drum with reverb creates a different effect in us than a dry snare sound with no effect.

Anything you do to any one particular sound in a mix will create a low-level dynamic. But it is not nearly as intense as the way that all the settings together can create an overall pattern that is much more dynamic and more critically important in the overall mix.

Level 2 – Overall Patterns

This is the pattern that is created by all the settings of a particular tool in the mix. For example, if you set all the volumes "even," with little variation between the loudest and softest sounds in the mix, versus setting the levels so that the volume range between the softest and loudest sounds is really wide, it will create a pattern of volumes that is a more intense dynamic than just setting one particular sound louder or softer.

If you create a lopsided mix with a vocal on one side and the rest of the band on the other side, versus a mix with balanced panning, it will affect the listener much more intensely than simply panning one sound left, right, or center.

An overall bright EQ for the entire mix, instead of a mellow bassy EQ, is way more critical to get right than the EQ of any one particular sound.

And, of course, if you create an overall mix with lots of effects, or none at all, it creates a much more noticeable dynamic than whether only one sound has an effect or not.

Level 3 – Movement (Changing Settings)

This is the movement created when you change settings during the mix, that is when you automate the volume, panning, EQ, or effects. It is, by far, the most intense of the three levels. If not appropriate for the song, it can overwhelm it, becoming the sole focus of attention at that moment. On the other hand, when such dynamics fit the song, it can create a whole new level of intensity, whereby the equipment is now functioning as a musical instrument in the mix. Therefore, it is good to always be on the lookout for songs in which this might be appropriate (and the band will let you do it).

Invisible Versus Visible Mixes

When used appropriately, Level 1 dynamics tend to create mixes that are invisible, or transparent. For example, when mixing big band music, acoustic jazz, or bluegrass, you should not hear the mix. The mix should be transparent to let the music show through.

However, in other styles of music, the mix might be totally visible creating its own dynamic. As mentioned, when utilizing Level 3 dynamics, the mix itself becomes a musical component of the song and must be performed in a way that works with the dynamics of the song. Pink Floyd has done this with surround sound concerts. Dance, electronica, and techno music also commonly utilize the mix as if it is another instrument in the song.

Dynamics Created by Studio Equipment Categorized by Emotional Effect

The key is to establish a connection between the technical equipment in the studio and the feelings and emotions found in music. The chart opposite will help you understand this concept. The middle column shows the tools. Columns two and four show the types of dynamics that can be created with the tools. And columns one and five show the types of feelings and emotions created by the dynamics.

Dynamics Created by Studio Equipment Categorized by Emotional Effect				
The key is to establish a connection between the technical equipment in the studio and the feelings and emotions found in music. The following chart will help you understand this concept. The middle column shows the tools. Columns two and four show the types of dynamics that can be created with the tools. Columns one and five show the types of feelings and emotions created by the dynamics.				
Calming Emotions				**Activating Emotions**
Ordered Structured Even Gothic Stable Normal Romantic	Even volume relationships with little variation between each sound and successive sounds.	**VOLUME**	Volume relationships that vary drastically between sounds and from section to section.	Interesting Exciting Wild Creative Crazy Fun New
Balanced Simple	Natural EQ between all instruments. All fit well together. As if you were there.	**EQUALIZATION**	Interesting or unique EQ in an instrument calls attention to it. Trippy overall EQ creates cool tension.	Unbalanced Complex
Positive Values				**Positive Values**
Warmth Peace Love Security Atmosphere Centeredness	Balanced, symmetrical placement. Placed in positions where they are separate from each other.	**PANNING**	Unbalanced, asymmetrical placement creates activation. Overlapping panning creates fullness.	Fun Creativity Catharsis Intrigue Perspective
	Dry, unaffected instruments that leave more "space" between the sounds, creating more clarity.	**EFFECTS**	Fattening, delays, flanging, chorusing, phasing all create more sounds making the mix bigger.	
Negative Values				**Negative Values**
Boredom Triteness Status Quo Commercial Simplistic	Use of compression creates more stability in volume. Riding the faders evens out the volumes even more.	**STABILIZATION VS MOVEMENT**	Changing (automated) volume, panning, EQ, and effects can create intense excitement.	Frenetic Chaotic Crazy Annoying Psychotic Abnormal Rock & Roll

Chart 11. Invisible Versus Visible Mixes

Now, let's take a closer look at the three levels of dynamics that can be created with each of the four control room tools – volume, EQ, panning, and effects. Here is a visual outline showing the three dynamics for each of the four tools.

Let's begin with volume.

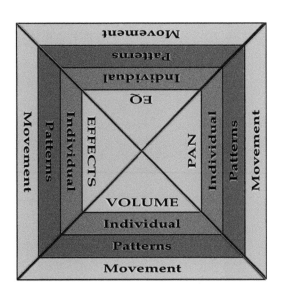

Visual 131.
Pyramid of Tools and Dynamics

Section A: Volume Control Dynamics

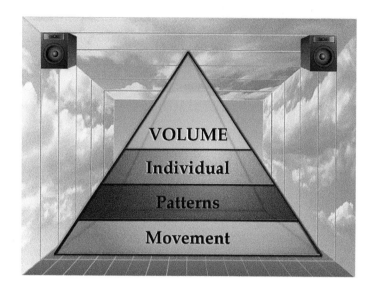

These are the musical and emotional dynamics that you can create by simply placing each instrument at different volume levels in a mix with faders or by moving the faders during the mix. Note that we are talking about the dynamics created by the engineer, not volume dynamics created by a musician playing an instrument louder or softer. Again, there are three levels. I'll begin with Level 1.

Fader Volume Levels

Level 1 Dynamics: Individual Volume Placement and Settings

You can create a wide range of emotional and musical dynamics, depending on how you set the faders on the console. The first and most basic level is based on where you place the volume of each sound in relation to the other sounds in the mix. As previously mentioned, if you place a vocal loud and out front in the mix, it will sound completely different than if it is placed back in the mix and softer.

The musical dynamics that can be created with volume placement are much more complex than most people imagine. Many people think of balancing the volume of sounds as making them even in volume. However, normally, you don't want all of the instruments to be the same volume. Many may be even in volume, however you usually want some instruments to be a bit louder than others, some in the foreground, some in the background, and some in between.

Each instrument is placed at a particular level in the mix based on the style of music, details within one of the thirteen aspects, and based on what the people involved want.

Every instrument has its own traditional volume level based on the style of music. Again, in many types of music, these levels have become strictly set. For example, the volume levels for big band, jazz, and even country have very little leeway. On the other hand, the set levels in rap, hip hop, and especially dance music are much looser. Although also much looser, even electronica and techno have developed some very strong traditions as to what levels each instrument should be placed at.

To explore these traditional levels in volume placement for each instrument, let's establish a scale for the volume levels that different instruments are placed at in a mix.

If we think of volume in decibels, based on sound pressure level, then a sound could be set at over 140 different volume levels in a mix. But in order to make this wide range of levels more manageable, I'll divide them into six different levels, or layers, from front to back. One is the loudest and six is the softest.

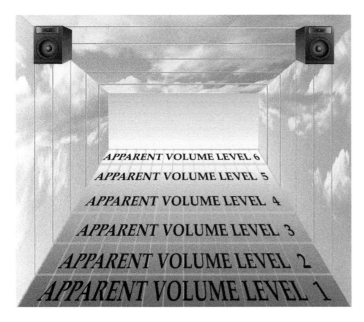

Visual 132.
Six Apparent Volume Levels

Remember that we are discussing relative apparent volumes. As previously discussed, the apparent volume of a sound is also dependent on the waveform of a sound. For example, a chainsaw sounds louder to the ear than a flute, even if they are both at the exact same volume. The apparent volume is the level that sounds seem to be to the ear.

1	Alarm Clocks	Explosions	Primal Screams			
2	Lead Vocals	Lead Inst's	Book	Horn Blasts	Symph Blasts	
3	Main Rhythm	Lead Vocal	Snare	Kick	Hi-Hat	Loud Effects
4	Rhythm Beds	Chordal Pads	Background Vocals	Strings	Reverb	
5	Distant Effects	Reverb	Hums	Background Vocals		
6	Subliminal	Whispers	Talking	Noises	Back Masking	

Chart 12. Six Ranges of Apparent Volume Levels with Typical Instruments in Each Level

Ranges of Apparent Volume Levels

I'll identify sounds that are commonly placed at each level so you can gain an understanding of the scale being used. Then I'll discuss each instrument sound and show how its volume varies, depending on the style of music, details of the song, and whims or expertise of the engineer and band.

Apparent Volume Level 1

Sounds at this volume are shockingly loud. In fact, it is quite rare and unusual to place sounds at this level. Commonly, only sounds that are very short in duration are this loud. If a sound is placed at this level for too long of a time, it often becomes annoying and dwarfs the rest of the mix. It would be thought of as either wrong or excessively creative. The alarm clocks in "Time" from *Dark Side of the Moon* by Pink Floyd are an example of interesting Level 1 sounds. Explosions, primal screams, and other special effects might also be this loud. These days on rare occasions a kick drum, snare, or bass might be at this level in hip hop or dance music.

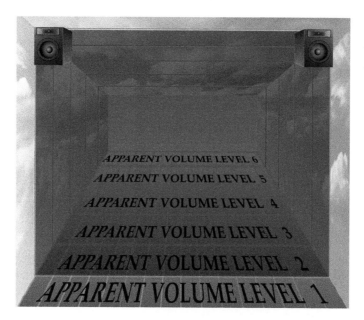

Visual 133.
Apparent Volume Level 1
Highlighted

Apparent Volume Level 2

The primary sounds at this volume are lead vocals and lead instruments. Rap, hip hop, and dance music commonly have the kick at this level (especially the 808 boom in rap). This level is actually thought of as quite loud in the mix, and is used for music in which the vocals or lyrics are the main focus of attention, such as big band and middle-of-the-road, or for vocals in pop music. In many types of rock and roll, the vocals are placed much lower in the mix. Horn blasts in big band music are often set at this level.

Besides the style of music dictating the level, a really great-sounding instrument (or voice) or a superb performance might cause an engineer to elevate a sound to this level of notoriety. Sometimes an engineer will place a sound at this level simply for its inherent shock value or to create more excitement.

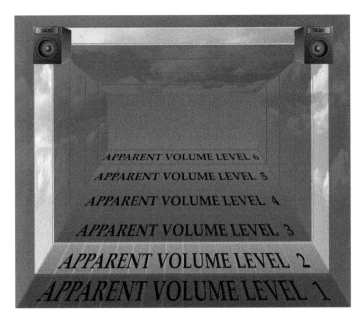

Visual 134.
Apparent Volume Level 2
Highlighted

Apparent Volume Level 3

Sounds at this level consist of primary rhythm parts, such as drums, bass, guitar, and keyboards. Lead vocals in a lot of rock and roll are also at this level when set back in the music. Other examples include kick drums in most rock and roll, snare drums in most dance music, and toms and cymbals in almost all styles of music. Hi-hat is only occasionally at this level, although jazz and dance music often place it here. Phil Collins was probably the first person to place reverb on the drums this loud.

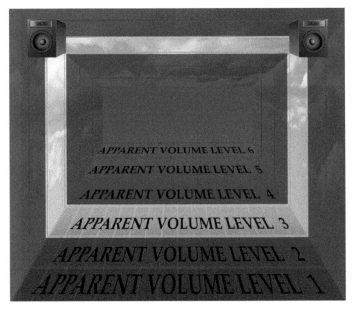

Visual 135.
Apparent Volume Level 3
Highlighted

Apparent Volume Level 4

Sounds at this volume include rhythm beds and chordal pads, such as background piano, keys, or guitar. Drums in lots of jazz, middle-of-the-road, and easy rock are often at this level. Background vocals, strings, and reverb are also often placed here.

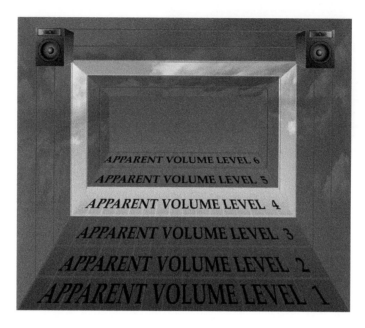

Visual 136.
Apparent Volume Level 4
Highlighted

Apparent Volume Level 5

Sounds at this low of a level are not very clear or distinct. They include the kick drum in jazz and big band music. Lots of effects and reverb are often placed here so that they can only be heard if you listen closely. Background vocals are sometimes relegated to this level. Other instruments are sometimes placed here only to fill in the mix.

Visual 137.
Apparent Volume Level 5
Highlighted

Apparent Volume Level 6

Sounds placed this far back in the mix are so soft that they are hard to detect. This also includes subliminal messages and back masking (sounds played backwards). Pink Floyd is well known for adding little whispers or almost subliminal sounds to draw you into the mix. Sounds at this level can be very effective, but it is important that they serve to add to the overall mix in some way. If these sounds do not fit just right, they might be perceived as noise.

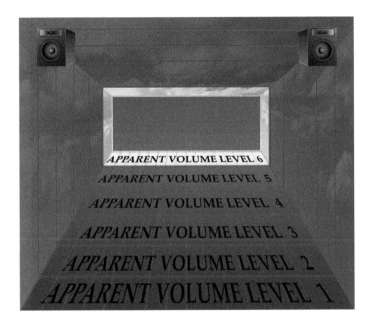

Visual 138.
Apparent Volume Level 6
Highlighted

These are the instruments that you most often find at each level. However, I have only placed them here to outline and establish the scale of six volume ranges so you can now use the scale to explore how actual instrument levels vary, depending on the style of music, song, and people involved.

Again, I'm not here to tell you the precise volume that sounds are placed in a mix.
I will only explain everything that an engineer takes into consideration
when trying to decide on the level of each sound in the mix.
Weigh the value of each consideration yourself, and
create the mix that you feel
is the most appropriate.

Vocals

Let's take a look at various examples of vocals placed at different levels in the mix. Depending on the style of music, the song, and how much the singer likes his or her own voice, lead vocals normally vary between Levels 2 and 4 (although a cappella music puts them at Level 1).

Apparent Volume Level 2

Vocals at this level are normally considered to be quite loud and out front compared to the rest of the band. Often the style of music dictates that the vocals are at this level. We commonly find vocals at Level 2 in opera and middle-of-the-road music, like Barry Manilow and Frank Sinatra. Most folk, big band, and country music also put the vocals right out front. Besides the style of music, the details of the song also affect the level placement. If the lyrics are the main focus in the music (Bob Dylan) or the singer is phenomenal (Janis Joplin, Celine Dion, Al Jarreau, and Bobby McFerrin), then the vocals might be brought way out front. Also, often the denser the arrangement, the lower the vocals are placed in the mix so that the details within the arrangement won't be dwarfed by the vocals. Loud vocals can make the rest of the mix sound wimpy. Generally, if a vocal part is extremely complex or detailed it will be turned up so that you are certain to catch the magic in the detail. For example, a rapper who is rappin' at the speed of light would be turned up just a bit louder in order to hear every little nuance.

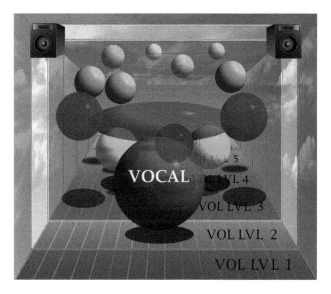

Visual 139.
Apparent Volume
Level 2 Vocals

Apparent Volume Level 3

Most vocals are mixed at Level 3: they are laid back in the mix but still loud enough to understand what's being said and hear the nuances in the vocal performance. Vocals at this level are not so loud that they dwarf the rest of the mix. Depending on the enunciation of the words, it might be more or less difficult to understand the words.

It is important to note that vocals often sound louder in the control room than they really are. This happens for a few reasons. First, studios often have very nice speakers that make all sounds clearer. Second, there is a phenomenon that happens whenever an instrument is soloed. If you solo a vocal, then take it off solo and listen to it in the mix, the vocal will sound louder than it really is because you just heard it in detail. Now that you know precisely what to listen for (sound, lyrics, and performance), it will seem louder in the mix. Also, vocals often sound louder when listening on small speakers that don't handle the low end very well. Finally, once you have learned the lyrics, the vocals will seem louder to you than they really are. Someone listening to a song for the first time will actually hear the vocals lower than you might hear them in the studio because they don't know the words.

Another consideration that some bands have figured out is that if you put the vocals lower in the mix, people often want to turn the music up louder. Take this tip as you like.

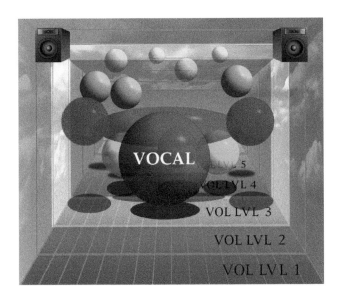

Visual 140.
Apparent Volume
Level 3 Vocals

Apparent Volume Level 4

Vocals at this level are so low that you normally can't understand the lyrics. As Mom used to say, "How can you understand what they're saying?" I would simply tell her, "You read the lyrics in the liner notes, Mom." A good amount of rock and roll, especially certain types of alternative rock, like Smashing Pumpkins, and Pearl Jam, have vocals this low in the mix. Pink Floyd, Enya, and Loreena McKennitt also sometimes have vocals at Level 4.

Besides the style of music, probably the most common reason for placing vocals at this level is so that they blend better with the music; therefore, they don't obscure the overall rhythm and melody of the song. The vocals become more of a melodic part of the music, so the lyrics are not so important.

I have also heard clients say that if the listener has to work harder to hear what the lyrics are saying, then they will have more meaning. Possibly true, if you can understand the lyrics at all. If you have lackluster lyrics or a vocal performance that could use some help, you normally don't put them right out front. But don't bury them in the mix too deep. Then you will not only have bad vocals, but once again, you will have a bad mix, too. Of course, with lyrics, don't forget – it takes all types for the world to go 'round. What you might consider bad lyrics might be creatively inspiring for other people (to a certain extent).

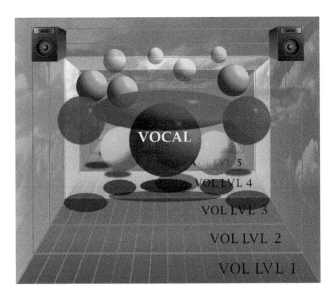

Visual 141.
Apparent Volume
Level 4 Vocals

For the most part, the style of music will be the primary consideration when deciding how loud a vocal should be in the mix. It will then be adjusted based on one of the thirteen aspects or the desires of the people in the room. Probably, the most common reason for vocals being at an inappropriate level has something to do with the singer – either they have a large ego or they feel very insecure. In my experience, the latter is the more common. "Turn me down (too much). I suck." The engineer for Jimi Hendrix, Eddie Kramer, said that Hendrix hated his vocals, and he had to work hard to get them up to an appropriate level.

Sometimes, as a band becomes famous and puts out subsequent albums, the vocals seem to be placed louder and more out front. This is a phenomenon that seems to happen in both hip hop and alternative rock. Often, I suspect, it is because they get hooked up with an experienced producer who makes sure they are clear and out front, as is the norm in the pop market.

Homework for the Rest of Your Life – Due (Do) Every Day

Every time you hear a song, notice at what level that the vocals have been placed in relation to the rest of the music.

Then, ask yourself "Why might they have put it there?" It is just game. Who knows for sure. What this does is helps you develop those real reasons that you can use on people when they say, "Please make the mix suck by turning me up too much."

Second, ask yourself, "Do I like the vocals at that level?" At first the answer might be, "I don't know . . . louder or softer . . . who cares?" But after checking out the volume of many vocals over time, you will develop very detailed values as to the volume that *you* feel is right.

Snare

The volume that a snare is placed in a mix is dependent on the style of music, the song, and the minds of the engineer and band members just as with vocals. The volume level of the snare, which seems to vary between Levels 2 and 5, has progressed up the scale over the years. Rock and roll was probably responsible for raising the level of the snare an entire level, then, in the 1960s, dance music and disco helped to raise the level of the snare to another level.

Apparent Volume Level 2

As with vocals, Level 2 is thought of as quite loud in the mix. Snare is normally placed at this level to whack people with excitement. Therefore, it is critical that the snare sound is really cool because such a loud snare can quickly become totally annoying. Really cool means unique, complex, or interesting without any accentuated irritating frequencies.

Some hip hop has the snare at this level (in some cases, practically Level 1). The snare is practically hurts (which can be fun). This is a perfect example of using the faders to create more excitement in a mix.

Various forms of rock (some Led Zeppelin and Bruce Springsteen) also have the snare this far out front in a mix. But don't confuse a large amount of reverb on a snare with being a loud snare.

Often a snare that was played loudly with lots of reverb is actually placed very low in volume. Also, a snare is usually only placed at this level when the tempo is slower, leaving more room in the mix. This is probably because, when it is this loud, it takes up a lot of space in the mix.

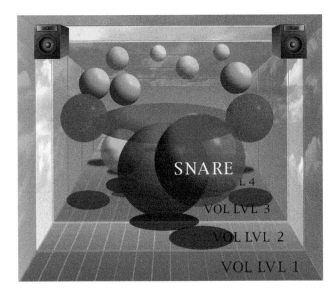

Visual 142.
Apparent Volume
Level 2 Snare

Apparent Volume Level 3

This level is most common for rock and roll. The snare is at this level for many styles of music, including heavy metal, blues, and now, even country.

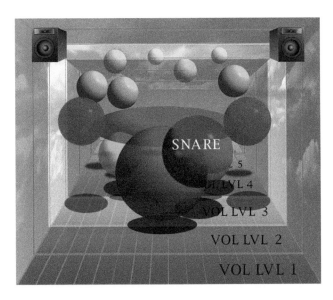

Visual 143.
Apparent Volume
Level 3 Snare

Apparent Volume Level 4

Big band, easy rock, new age, and 1950s and 1960s rock music often have the snare mixed this low. Most ballads will place the snare at this level, though there are some ballads that have a massive snare as loud as Level 2.

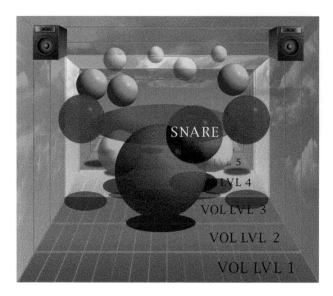

Visual 144.
Apparent Volume
Level 4 Snare

Apparent Volume Level 5

Big band music and acoustic jazz sometime have the snare this low.

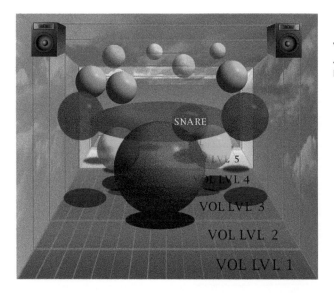

Visual 145.
Apparent Volume
Level 5 Snare

Some general rules include: the better the snare sound, the louder it is placed in the mix; the slower the tempo, the louder the snare; and the busier the arrangement, the lower the snare. These rules are not that strict . . . just good pointers to consider. But remember the best mix is one that fits the details of the song. It enhances the song more, making it more powerful.

Homework for the Rest of Your Life – Due (Do) Every Day

Now, whenever you hear a song (you might even put one on now), check out and define the level of the snare drum. And ask yourself why they might have put it at that volume, and do you like it.

Kick Drum

Again, the style of music will be the biggest influence in the volume in the mix. Kick drums tend to span Levels 2–5. Kick drums have been climbing the volume scale throughout history, probably suppressed for years because it was considered to be the beat of the devil. Of course, rock brought it up one level. Then, heavy metal was responsible for raising it another level. Then, rap and hip hop came along and put it off the scale. Now, you find the kick drum at extremely loud levels in all kinds of modern music.

The density of the arrangement tends to be the primary component of the thirteen aspects that affects the level of the kick drum. Since a kick takes up so much space in the mix with all of its low frequencies, it is important that you don't use up all of your space between the speakers by turning it up too loud. The sparser the mix, the more room there is. In a busy mix it can easily start masking other sounds. Be careful.

Of course, a really cool sounding kick drum and a great performance might elicit a little nudge in volume.

Apparent Volume Level 1

Rarely is a kick drum ever this loud; however, if you think of the 808 rap boom as a kick drum, then you'll sometimes see it placed here in rap and some dance music, such as "deep house."

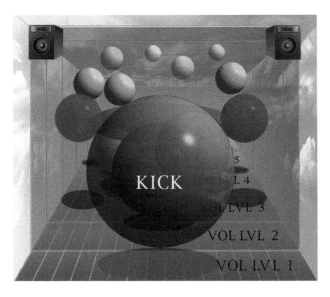

Visual 146.
Apparent Volume
Level 1 Kick

Apparent Volume Level 2

Rap booms are at this level, as is the primary kick in hip hop and house music. The kick drum in heavy metal is sometimes at this level, though it is usually raised to this level for only a short period of time in certain sections of the song. Occasionally a ballad will even have the kick at this level, and engineers have been known to make the kick this loud in blues and reggae.

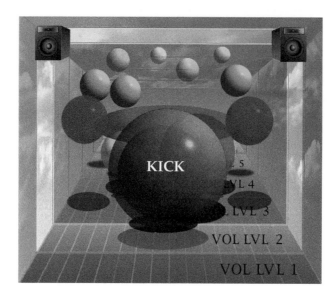

Visual 147.
Apparent Volume
Level 2 Kick

Apparent Volume Level 3

This is the most common level for the kick in most styles of music, especially rock, blues, jazz, and country.

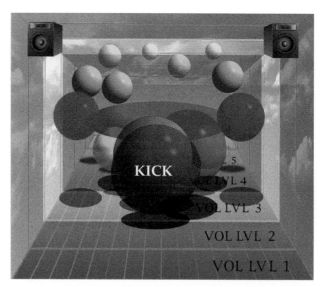

Visual 148.
Apparent Volume
Level 3 Kick

Apparent Volume Level 4

Jazz and new age, as well as a lot of ballads, commonly have the kick at this level. It is interesting that much of Jimi Hendrix's music was mixed with the kick drum down around Level 4 so that you could hardly hear it. Of course, this was common in many songs and styles of music in the 1960s.

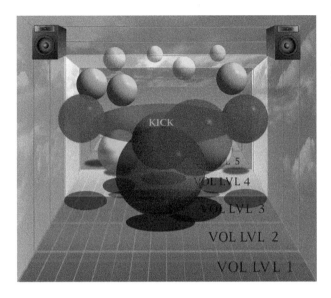

Visual 149.
Apparent Volume
Level 4 Kick

Apparent Volume Level 5

Big band music commonly has the level of the kick down this low in the mix.

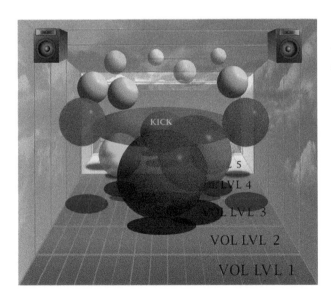

Visual 150.
Apparent Volume
Level 5 Kick

Just like the snare drum, the kick drum volume is primarily based on the style of music. However, both the song and the particular instrument sounds also contribute to this decision. The more interesting and complex the kick sound, the louder it usually is in the mix. The slower the tempo, the louder the kick. The busier the arrangement, the lower the kick.

Homework for the Rest of Your Life – Due (Do) Every Day

Now, as before, check out the kick drum level in every song you listen to – for the rest of your life. And ask yourself why they might have put it at that volume, and do you like it.

Bass Guitar

The bass guitar normally spans Levels 1–4. As usual, the style of music is the primary determining factor in deciding on the level of the bass. With the evolution of rock and then disco, the bass has crept up the volume scale. Then rap began a revolution that not only helped raise the level of bass guitars in mixes, but resulted in a change in the hardware we use. When you go to a stereo store, you see things like Mega Bass and boom boxes. And normal home stereos are now capable of handling a lot more bass than before.

Of the thirteen aspects, again the density of the mix plays an important role in the volume placement for bass sounds. Because it takes up so much space in a mix, it is often placed lower in the mix, so it doesn't mask the other instruments too much. However, the more room in the mix, the more it is commonly brought up to fill out the mix. Once again, the quality of the performance and sound quality will often affect the volume to a small degree.

And, of course, there is always the bass players opinion to deal with.

Apparent Volume Level 1

It is quite rare that the bass is this loud, although there is some rap, hip hop, and house music that have gone off the deep end. Ordinarily, the bass is only turned up to this level for a moment in a special section of the song.

Visual 151.
Apparent Volume
Level 1 Bass

Apparent Volume Level 2

Reggae and the blues often have the bass this loud. Because the bass is carrying the song in blues, it is often needed to help fill out the sparse arrangements commonly found in these mixes. When the bass is a lead part in the song or music, it is often right out front at this level. This is often the case in jazz, especially if the bass guitar is a fretless. Primus and Stanley Clarke are good examples of the bass played at this level.

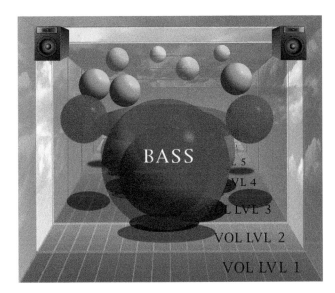

Visual 152.
Apparent Volume
Level 2 Bass

Apparent Volume Level 3

This is the most common level for the bass guitar for most styles of music: not so loud that it takes up too much space, but loud enough to still be heard well.

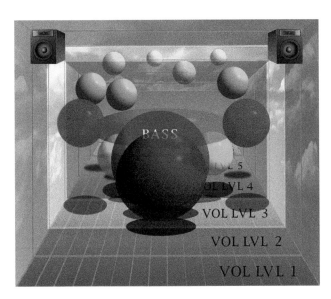

Visual 153.
Apparent Volume
Level 3 Bass

Apparent Volume Level 4

The bass guitar in a good amount of rock and roll is down at this level, so it doesn't get in the way. In big band music, you often find the bass here. In fact, when you have a standup or acoustic bass, it often ends up here.

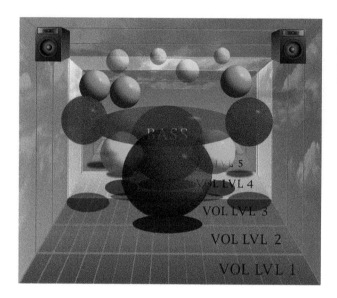

Visual 154.
Apparent Volume
Level 4 Bass

Commonly, the fewer instruments in a mix, the louder the bass because you need something to fill out the space between the speakers. Also, if you have a lot of instruments, there just isn't enough room for the bass guitar, and it will mask the other sounds if too loud.

Homework for the Rest of Your Life – Due (Do) Every Day

What level is the bass in your favorite style of music? Check it out, whenever your ears are open. And ask yourself why they might have put it at that volume, and do you like it.

Tom Toms

Toms span the entire volume scale from Levels 1 to 6. Although dependent on the style of music (normally you don't have really loud toms in new age music), the details of the song and the preferences of the engineer and band seem to be the most common determinants of tom levels.

When sounds don't last very long, they can be turned up louder: they're gone before you know it. Because of this, toms are often boosted more than you would think. Generally, the more they are played in the song, the lower toms tend to be placed in the mix. Also note, the brightness of the toms in the mix makes a huge difference as to how much they are masked by the rest of the mix.

Apparent Volume Level 2

Toms are sometimes placed this loud because their duration is so short, and they are played so sparsely.

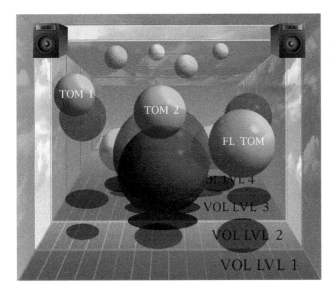

Visual 155.
Apparent Volume
Level 2 Toms

Apparent Volume Level 3

This is the most common level for toms in most styles of music – present, but not so loud that they break up the beat or rhythm of the song too much.

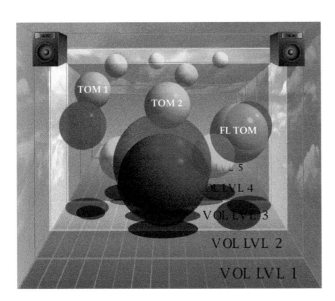

Visual 156.
Apparent Volume
Level 3 Toms

Apparent Volume Level 4

Toms are normally not very loud in most types of music, somewhere around Level 4. The most common reason for putting toms at this level is so they don't interrupt the flow of the song's rhythm too much. However, sometimes I think engineers don't turn the toms up very loud because it makes the cymbals sound so horrendous. This has to do with the problem of cymbals bleeding into the tom mics. When this happens, the sound of the cymbals in the tom mics is irritating because the sound is reflecting off the tom heads, especially if the toms need to be brightened a lot with EQ.

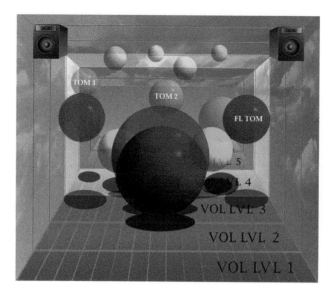

Visual 157.
Apparent Volume
Level 4 Toms

Apparent Volume Levels 5 and 6

I wouldn't doubt that engineers who mix toms at Levels 5 or 6 either don't like the tom parts, don't like toms in the first place, or forgot about them.

Homework for the Rest of Your Life – Due (Do) Every Day

Don't forget to gauge the level of the toms in every mix you hear from now on. And ask yourself why they might have put it at that volume, and do you like it.

Hi-Hat

The hi-hat normally fluctuates between Levels 2 and 5. The level of the hi-hat depends mostly on the style of music, although the details of the song often make a big difference, too. Once again, the quality of the performance and the sound of the hi-hat can result in a minor adjustment in volume.

Apparent Volume Level 2

Hi-hats are normally the loudest in heavy metal and R&B music. Hip hop, jazz, and dance music often place it at this level as well.

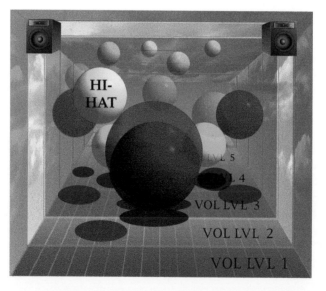

Visual 158.
Apparent Volume
Level 2 Hi-Hat

Apparent Volume Levels 3 and 4

Hi-hats commonly fluctuate between these two levels for most styles of music, especially rock and roll.

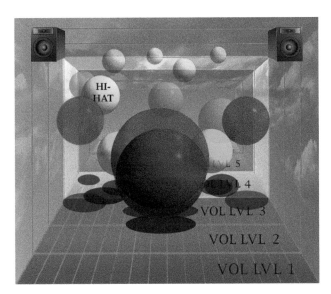

Visual 159.
Apparent Volume
Level 4 Hi-Hat

Apparent Volume Level 5

Although the hi-hat doesn't take up much space in a mix, it does cut through well. Not only is it an edgy type of sound, but it also resides in a frequency range with very few other instruments. Therefore, even when placed low in the mix, it can often still be heard just fine.

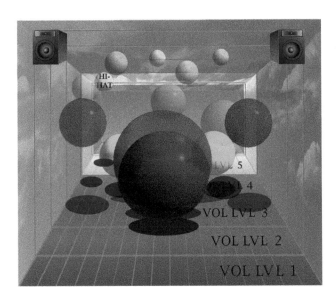

Visual 160.
Apparent Volume
Level 5 Hi-Hat

Homework for the Rest of Your Life – Due (Do) Every Day

Keep a look out for the precise level that hi-hats are placed in mixes, especially for different styles of music. And ask yourself why they might have put it at that volume, and do you like it.

Cymbals

Cymbals range the entire gamut, from Level 1 through 6. The style of music makes a difference, but even more importantly, the particular sound of the cymbals and the parts being played in the song tend to affect the level the most. However, since the musical traditions are not very strict, often the preferences of the engineer and the band affect the final level.

Apparent Volume Level 2

It isn't too often that cymbals are at this level, although Led Zeppelin and Creedence Clearwater Revival placed them this loud occasionally.

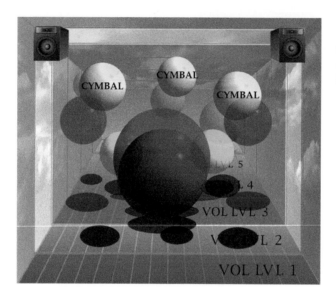

Visual 161.
Apparent Volume
Level 2 Cymbals

Apparent Volume Levels 3 and 4

Most cymbals are set at these levels, so that they are evident but still blend in with the rest of the instruments in the song.

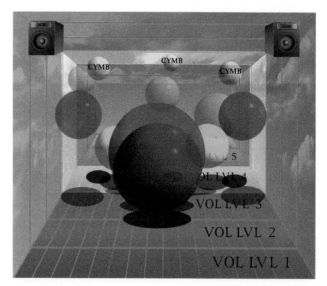

Visual 162.
Apparent Volume
Level 4 Cymbals

Cymbals at Level 5 or 6 are often there because of their sound or masking from other sounds in the mix.

Homework for the Rest of Your Life – Due (Do) Every Day

Now pay attention to cymbal levels in songs on the radio and CDs. And ask yourself why they might have put it at that volume, and do you like it.

Effects

The volume of different effects varies widely over the level spectrum. As effects have become more commonplace in our society, the volume for each has progressed up the scale over the years.

There are a few traditions as to the amount or type of effect that is placed on any particular instrument. The most common tradition for reverb is that it is placed on the snare and vocals. However, hip hop rarely has much reverb on either, these days. The snare is commonly dry in jazz, but it varies in rock and roll. Of course, classical and big band music don't have effects other than a slight amount of reverb. It is against the law. As previously mentioned, use effects on this style of music, and you *will* go to jail.

Use of delays is often based on the style of music also. Commonly, a long delay with feedback is brought up at the end of a line, line, line. The volume of delays varies tremendously based on details of the song.

Flanging, chorusing, and phasing volume levels are set based on the style and details of the song. Flanging is commonly used on guitars and keyboards. It is only occasionally used on vocals. Chorusing is often heard on vocals, background vocals, guitars (including bass guitar), and keyboards. Phasing is heard on a wide range of instruments, except for percussion, but is much less common overall.

Apparent Volume Level 1

Effects are rarely this loud in a mix, usually only if they are extremely short in duration. When placed at this level, they can be shocking enough to have a lasting effect on the listener, longer than the duration of the sound itself.

Visual 163.
Apparent Volume
Level 1 Effects

Apparent Volume Level 2

Reverb is occasionally this loud on a snare drum, depending on the song and style of music. When a delay is this loud, it is as loud or louder than the sound it is on. For example, a lead guitar or vocal might easily have a delay this loud. Flange-type effects are also sometimes at this level.

Visual 164.
Apparent Volume
Level 2 Effects

Apparent Volume Levels 3 and 4

Most effects are placed at this level – loud enough to hear the detail within the effect, but not so loud that it overwhelms other sounds in the mix.

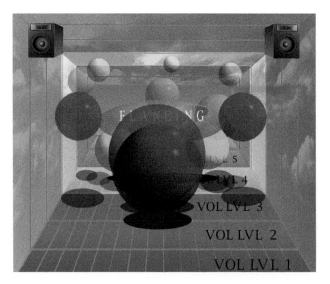

Visual 165.
Apparent Volume
Level 4 Effects

Apparent Volume Level 5

Reverb is often at this level and is quite unnoticeable to most listeners.

Visual 166.
Apparent Volume
Level 5 Effects

Other Instruments

I have covered only the most common instruments found in recordings. There are, of course, a huge number of other instruments. *Note the level of each and every instrument in the music you listen to.*

Summary

As you can see, there is an incredible variety of musical dynamics based on the level at which you set each sound in the mix. This is important because so many bands want their mixes to sound like they traditionally do for their style of music.

When bands complain that the mix doesn't sound quite right, it is often as simple as an instrument being placed at the wrong volume. Anyone who has grown up listening to or playing a particular style of music often has these traditional levels in their heart. They can feel when the level of any instrument is not the way it is supposed to be, but *they rarely know why*. Inexperienced engineers will often go on a wild goose chase, changing EQs, and effects to try and resolve the problem, when it might actually be that the rhythm guitar is too loud compared to the vocals, or the kick drum is the wrong volume compared to the bass guitar. Therefore, it is critical to study these traditional volume levels for each instrument in each style of music. This doesn't mean that there is no room for creativity, but you must be creative within the limits of tradition if the band wants it to sound like their style of music. *And most bands do!*

It is normally important that the mix fits the style of music, but each one of the thirteen aspects might also make its own contribution. The concept of a song about someone that is angry and loud might appropriately have the vocal too loud – at Level 1 even. A simple melody or rhythm might make you turn the volume down a tiny bit; whereas a busy melody or rhythm is often bumped up ever so slightly. Beautiful harmonies might be highlighted with a few decibels of boost. An unusual song structure might have a sound that is placed low in the mix in order to accentuate a loud sound in the next section. A very busy arrangement might cause you to lower a sound back into the mix. A really unique and intriguing sound might be highlighted with a little extra volume. And, of course, a really great shining performance on any instrument should not be missed by being placed too low in the mix. An instrument with a bad buzz in it, or some type of distortion that can't be resolved, will be turned down a bit, of course. And finally, you might turn down a sound that has an extra boost of high-frequency equalization or a lot of effects in order to compensate. Even the mix can affect the mix.

> Together,
> the style of music and
> the thirteen aspects
> are telling you
> precisely where the volume should be.

The wrench in the whole matter is often the people involved. The best mixes commonly are a result of everyone in the room focusing on what the song is telling you to do with the volume of each and every sound! It's when you get someone whose ego or insecurity starts guiding the mix that you run into trouble. But as previously covered, sometimes people do receive inspirations from God (or who knows where?). So the trick is to listen to whether the ideas fit – or not – in every case.

> Once you learn how to differentiate between six levels of volume, the next step is to learn to hear finer and finer levels of volume, so that you can see more than just six levels: twelve levels is cool, but twenty-four is incredible.

Homework for the Rest of Your Life – Due (Do) Every Day

Whenever you have time to listen, determine the relative volume level of each sound in the song (on a scale of 1–6).

Here are some common reasons why instruments are placed at specific volumes:
1. That's the way it is normally done in this style of music.
2. Something about one of the thirteen aspects caused the engineer to place each sound at that level. The most common are:
    a. Great performance
    b. Great sound
    c. Great lyrics or music
3. Somebody in the room asked for the sound to be at that level.
4. The engineer was either nuts or on drugs.

Again, this exercise helps you to develop your repertoire of reasons to help explain to people why things should be the way they should be.

Asking yourself, "Do I like the volume level of each sound in the mix?" helps to develop your own values. After a few months (or few years), you will know exactly where you like the levels of different instrument sounds for various styles of music and songs. Then, when you go into the studio, you'll no longer be unsure about any volume placement. You'll not only have a perspective on where other engineers commonly place sounds, you'll know exactly where you like the volume of each sound. Then you are on your way to becoming a real recording engineer.

Level 2 Dynamics: Overall Patterns of Volume

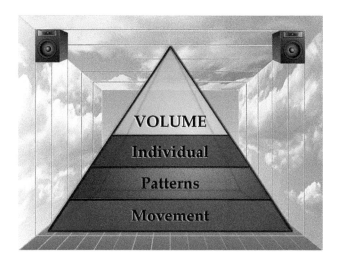

Compared to the individual volume levels of an instrument, overall patterns of volume are much more important in a mix than any one single sound being placed louder or softer. They also create a stronger emotional dynamic. In other words, they affect us more.

Certain styles of music have developed their own traditional levels of how even or uneven the overall volumes are set. And again, certain styles of music have stricter rules than others. It is important to get to know these traditional levels, so you can push the limits of creativity and change the world.

In some styles of mixes, the volumes are set evenly so that there is very little variation between the loudest and softest sounds. New age music, alternative rock (REM, Smashing Pumpkins, Nine Inch Nails, U2, etc.), middle-of-the-road music, country music, and easy rock are often mixed with very little volume variation. You could say that muzak is the extreme example. Even volumes might also be appropriate for a love song.

Visual 167.
Song with Even Volumes

Sometimes, it is necessary to compress sounds a bit more to help make the volumes more even. However, you can only compress things so much before they sound squashed. Often, it becomes necessary to "ride" (move) the faders up and down in volume to keep things even; or use "line automation" and draw in volume changes to mitigate the dynamics of the instruments. However, because this actually means moving the faders, I will deal with this more in the next section, "Changing Volume Levels."

Alternatively, some styles of music are mixed with extreme variations between the softest and loudest sounds, like this:

Visual 168.
Song with Uneven Volumes

Lots of rock and roll, dance music, and rap are mixed this way. Big band music is also a perfect example of this type of mix. You might have extremely soft sounds followed by huge horn blasts. Even some classical music is this dynamic. Pink Floyd is well known for trying to shock you with alarm clocks and explosions. It can be quite fun and exhilarating.

Whether the mix is even or uneven is mostly based on the style of music. However, the type of song also helps to determine the overall evenness of volumes. For example, a ballad might be mixed with volumes that are more equal to each other to preserve the overall mellow feel. On the other hand, a song about "trauma in life" might very well have some shocking volume differences. It is helpful to listen to the details of the song to help you determine whether levels should be even or uneven.

Besides the style of music and the type of song, the density of the arrangement will also affect how even the mix is. Generally (but not always), the denser the arrangement, the wider the volume range, from softest to loudest sound in the mix. If you have a sparse mix, it is more common to create an even mix so the sounds appear more cohesive.

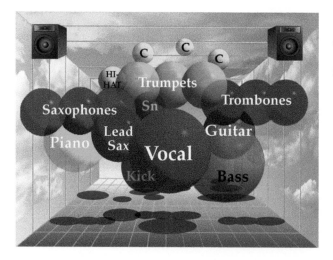

Visual 169.
Uneven Volumes:
Big Band Mix

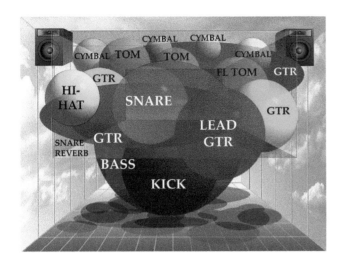

Visual 170.
Even Volumes:
Heavy Metal Mix

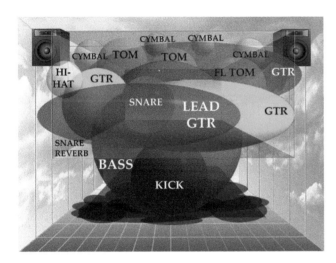

Visual 171.
Even Volumes:
Alternative Rock Mix

Mixes with even volumes are much harder to perform and much more difficult to get right because you are under the microscope. Making a change from one level to another is an extremely small change on the fader. Therefore, you have to be very detailed and pay attention to minute differences in volume.

On the other hand, if you are creating a mix with uneven volumes, it's easy – the difference between levels is so great, you have a lot more leeway.

Level 3 Dynamics: Changing Volume Levels

Certain styles of music do not allow volume changes – especially volume changes that can be perceived as noticeable engineering moves. With folk, bluegrass, acoustic jazz, and other natural styles of music . . . *move the fader, and you're in trouble.* Once again, the style of music often dictates whether you can do fader moves that act as a musical component in the mix.

Meanwhile, there are many subtle fader moves that can even contribute to a mix that is perceived as more stable and more traditional. Obviously, a lead instrument should be turned up if it is on the same track as a rhythm part, then brought back down after the lead is over.

Generally, volume changes made at the beginning of a song section are not as noticeable as changes made in the middle of a song section. If the level of a sound is changed at a good transition point, such as the beginning of a chorus or lead break, the dynamic created is often subtle.

If a volume is changed in the middle of a song section, it creates such a strong dynamic that it causes the listener to focus on the change you are making. Therefore, it should be executed as if it were a musical performance – that is, in time with the tempo or other changes that are occurring in the song.

The biggest volume dynamic is when the volume of the entire mix is raised or lowered. The master stereo fader volumes are not commonly changed except at the beginning or end of the song. Fading in the entire mix at the beginning of a song creates a very nice and smooth dynamic; the Beatles used this on "Eight Days a Week." I've also heard songs where the entire mix is faded out, faded back in, faded out again, and faded in one more time. A very cool effect is to cut, boost, or gradually fade the overall volume in the middle of a song. Such dynamics can be quite effective. Fading out a particular section of instruments (such as drums) and then fading it back in can also serve to wake people up.

In addition to moving a fader to create volume dynamics, you will usually need to adjust levels to keep the volumes even. Compressor/limiters can only do so much before they make a sound unnatural. Therefore, you can create another dynamic – to actually even out volumes more – by riding the faders throughout the mix or using line automation. It's actually a skill that you can learn very quickly. You listen very closely and the moment you hear something getting too loud, immediately bring the volume down. When it gets too soft, immediately bump up the volume. This is especially effective when you have automation that records your moves, and you can make minor adjustments in case you don't get it quite right.

Magic Trick

There is an interesting phenomenon that you can use to create a clear mix out of an extremely busy and dense arrangement. It works especially well with multiple musical parts that are repeating melodic lines. Start by turning up the first repeating melodic line so you can hear it clearly. Now, bring it down so it is low in the mix. Even if you can barely hear the sound, you now know precisely what the sound is and what it is doing because you just heard it highlighted. In your mind, the sound and musical part is still perfectly clear. You can do this with each repeating melodic line until you have four or five sounds placed low in the mix where they can hardly be heard; however, in the listener's head, all the sounds and musical parts are perfectly clear. You have a perfectly clear mix, even though you can hardly hear the sounds at all. However, if someone were to walk into the room in the middle of the song, they would not agree.

Although changing levels in a mix can create a major dynamic, you can create much more subtle (and often more effective) dynamics by making minor volume changes in various sections of a song. For example, you might boost the volume of the guitars (ever so slightly) in the chorus, raise the snare and snare reverb ever so slightly in the lead break, then bring up the bass guitar and kick drum (again, just a touch) in the vamp at the end of the song. These subtle volume changes can add serious magic to your mix.

These three levels of dynamics –

individual volume settings,

overall volume patterns, and

volume movement –

make up all that can be done with volume faders in a mix.

Compressor/Limiters

Just as volume faders can create a wide range of dynamics, compressor/limiters can also be used to create musical dynamics to fit the music or song. Compressor/limiters are often used for technical reasons, such as to get a better signal-to-noise ratio (less hiss) or just to be able to record hotter on the tape (or hard disk). However, this section covers how they are used to create a musical or emotional component that fits the emotions in the music and song.

Level 1 Dynamics: Individual Compressor/Limiter Placement and Settings

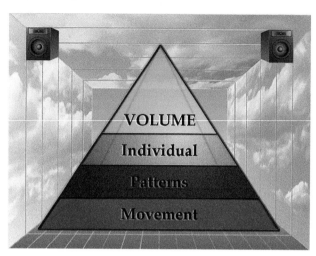

Sounds are compressed based on the dynamic range of the sound itself. For example, a "screamer"-type vocal (Aretha Franklin, Axl Rose, Janis Joplin, and Pavarotti) will normally be compressed more to account for the huge difference in volume from soft to loud. However, there are certain sounds that are compressed more out of traditions that have developed over years of recording and mixing.

First, most acoustic instrument sounds are compressed to even out the volume of the playing and to smooth out any resonances in the instrument. Vocals and bass guitar are almost always compressed. Many engineers will compress the kick drum for presence, although there are some who do not believe in compressing the kick. If the drummer is good and has control of the volume of each kick lick, then compression may not be necessary. However, compression does make the attack of the kick drum sharper.

The density of the arrangement also affects the amount of compression. Many instruments are only compressed when they are placed in a mix (as opposed to being solo). For example, it is rare for anyone to compress a solo piano; however, pianos are commonly compressed when placed in a mix, especially a busy one. Acoustic guitars are also commonly compressed in a mix. In fact, as mentioned before, the busier the mix, the more the individual sounds are compressed. This is done in order to minimize the huge amount of movement by the natural fluctuations in volume of each sound. We can only handle so much stimuli before we start to lose it.

It is also common to compress the loudest and softest sounds in a mix more than other sounds. By compressing a loud sound, you can make it even more present and in your face. As mentioned, this is also done in radio and TV commercials, but for a vocal or lead instrument it is desirable. Lead guitars are often compressed more, simply because some people like the intensity and power of an electric guitar right in their face. Remember, one of the primary functions of a compressor/limiter is to make a sound more present. Likewise, if a sound is placed low in a mix, and it is fluctuating in volume (as all sounds normally do – just look at a VU meter), the soft parts of the sound are easily lost in the mix as they get masked. By stabilizing the volume of the sound, you can place it in the background, and it will still be present!

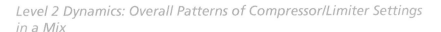

Level 2 Dynamics: Overall Patterns of Compressor/Limiter Settings in a Mix

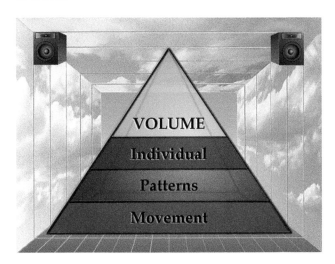

The overall amount of compression on a mix is obviously more noticeable than any individual setting. There are two ways that it can be applied. The first is based on the combined overall amount of compression added to each sound individually because some sounds may have no compression at all. The second is based on the amount of compression added to the overall mix once it has been mixed. This process, commonly done in mastering for CD pressing, only compresses the loudest sounds in the entire mix. When sounds are compressed individually during the recording and mixing of the song, all volumes can be compressed, not just the loudest ones.

There is a common compression technique that is used on a drum kit recorded live in the studio. First, you make a sub-mix of the drums to two tracks, then place a stereo compressor on the drum mix. You then mix this compressed stereo track back in with the original drums. Although you normally don't want the original signal when compressing, in this case, it makes the entire drum mix extremely present and fatter at the same time.

Regardless of the way overall compression is applied, or calculated, certain styles of music have developed quite strong traditions as to how much they are compressed. For example, most pop music has more overall compression than most country music or punk. This can be perceived as "polish," which some people complain is part of being over-produced. You can see the amount of overall compression on the mixdown VU meters. The meters barely move on highly compressed material. As previously discussed, overall compression and limiting is part of mastering.

Rhythm and blues and middle-of-the-road music are often compressed more than other styles. Acoustic music, such as bluegrass and acoustic jazz, are commonly not compressed as much. Again, these rules are made to be broken. Much of the electronic type of music (anything that uses a lot of synthesizers and drum machines) will often sound more compressed because many synthesized sounds have been compressed previous to being placed in the synth. Therefore, much of the hip hop and dance music mixes sound highly compressed. Many engineers feel that you should not compress sound healing music as much in order to make it more natural. I feel the opposite. I actually compress it a good amount so that no sounds jump and shock you – destroying the healing effect.

Homework for the Rest of Your Life – Due (Do) Every Day

Pay attention to the overall amount of compression that seems to be going on in each song you hear, and develop your own values for how much compression you like.

*Level 3 Dynamics: Changing Compressor/Limiter Settings
(Levels and Parameters)*

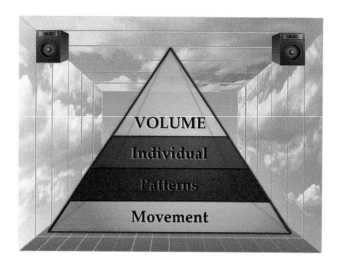

Changing the amount of compression, ratio setting, attack, or release time is rare in the midst of a mix. It is commonly done when mixing voices in a movie, video, or commercial. Narration is often compressed more than dialog, for example. Because compression has the effect of making something sound more present, it can actually be used to create a dynamic that seems to move from more distant to more present or vice versa. One of the most dynamic effects is to change from no compression to limiting. This makes the sound seem to jump right out at you.

Normally, changes in settings are done at musical transitions in the song – such as at the beginning of a verse, chorus, bridge, or lead break – so that the change is not so abrupt or shocking. However, it just might be an interesting effect (if appropriate) to change the settings of compression while in the middle of a vocal or lead solo. When doing this, you are creating a dynamic effect so strong that it will show through the mix; therefore, it should be musically performed so that it fits the song.

Noise Gates

Level 1 Dynamics: Individual Noise Gate Placement and Relative Settings

The use of noise gates is primarily based on technical considerations: that is, getting rid of low volume noises and bleed from other instruments in the room. The only consideration for using a noise gate that might have developed any sort of tradition would be using a noise gate to shorten the duration of a sound by chopping off the attack or release. This is not to say that it isn't a cool effect and should not be utilized; it just isn't very common.

Level 2 Dynamics: Patterns of Noise Gate Placement

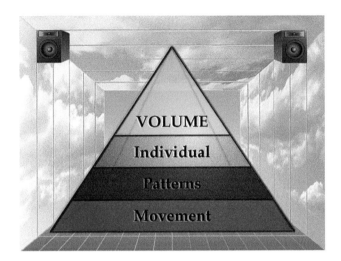

Extensive use of noise gates in a mix primarily results in more precise imaging between the speakers because of the way that noise gates help to isolate a sound and get rid of phase cancellation. It seems that most pop music is mixed using gates. Styles of music that are more focused on clarity, such as Steely Dan, will often use more noise gates overall.

On the other hand, some engineers prefer not to use noise gates much, especially on the drums. Using less noise gates will help make the mix sound more "live" in some engineer's eyes (or ears).

Level 3 Dynamics: Changing Noise Gate Settings (Levels and Parameters)
Settings on noise gates are rarely changed during a mix. However, to do so may give the illusion of a mix that is becoming more and more clean, with more precise imaging. You could also use noise gates to shorten the duration of a sound bit by bit, which might be totally appropriate for a song about losing weight or a shrinking reality.

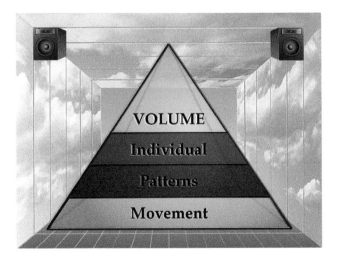

Section B: Equalization Dynamics

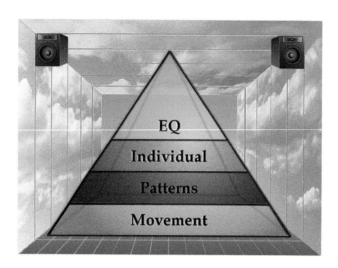

If you have been doing much recording, you know that the limits of creativity are tight with EQ. There is very little room for creativity. With volume, you have more freedom. With EQ, you often work hard just to get things to sound "right." If you are going to be creative with EQ, you have little leeway.

What we consider to be natural EQ for each instrument within each style of music has become entrenched in our audio consciousness. The brightness, midrange, and fullness of each instrument are now strictly defined. In fact, if you don't EQ the instruments based on these traditions, it is either considered to be wrong or exceedingly creative. But it's not only tradition. Our psyches have little patience for wayward frequencies jumping out and attacking us.

Because creativity is so limited, it is important to look closer – refine your focus, so to speak. It's like looking at the frequencies with a magnifying glass. Once you "zoom in" on the exact limitations of good and bad EQ, you can push the limits. You can be creative without going off the deep end. Often this means minuscule changes – it doesn't take much to screw up EQ. Of course, you can only be creative with EQ when it is appropriate for the style of music, the song, and if the band will let you.

As with volume, there are three levels of dynamics that can be created with equalization. First, the individual brightness, midrange, and bassiness of each instrument (relative to the rest) create small but definite differences. Of course, each instrument has developed its own traditions for what we consider to be normal EQ. If we set the EQ different from these traditions, we are creating a unique dynamic that affects the overall perception of the music or song. There is a much stronger dynamic created by the combination of all of the relative EQ settings together in the song. But the most powerful dynamic you can create with an equalizer is to change the EQ during the song. This tends to be even more intense than changing volume levels during a song.

Equalization Dynamics

Level 1 Dynamics: Individual Equalization Settings and Placement

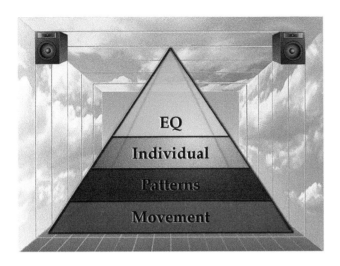

There are two primary ways in which the individual EQ of a sound can create musical and emotional dynamics. First, the individual EQ of a sound can either be made to be "natural" or "interesting."

Natural EQ

In the beginning, the basic goal of using EQ was to make the sound natural – just like it sounded in the room where the instrument was. You can't get any more natural than that, right? The only problem is that natural is not natural any more. These days natural is defined by what is currently on CDs and the radio. We have become addicted to crisper, brighter, and cleaner, as well as fatter, fuller, and bigger. Therefore, to make a sound natural can be boring and unnaturally dull by today's standards. What we hear on the radio and on CDs these days is much brighter, crisper, and bassier than the real thing. If it isn't bright enough and doesn't have enough low end, it won't be considered right.

Be careful when using EQ to move sounds up or down to clear out a space for a sound or when trying to "separate" two sounds from each other. This problem often happens when the engineer tries to use EQ to fix an arrangement problem when too many instruments are located in the same frequency range. EQ will work, but it will very quickly make your sound unnatural. For example, you might turn up the highs on one guitar, while turning up the midrange on another guitar – this separates the sounds, making them a little more present and discernible. The problem is that when you listen to the mix the next day, the guitars don't sound natural: one sounds too bright and the other sounds too midrangey. It is really easy to be seduced by the clarity you have achieved and forget to notice that your instrument no longer sounds natural. Therefore, it is always a good idea to double-check that the sound is still natural when in solo. You might find that you should compromise if the sound is unnatural. This will help you end up with an overall EQ that sounds natural and has some separation.

The following is a list of a few of the most common instruments and their typical EQs, to give you an idea of what we have come to expect based on the style of music and song. Of course, these EQ settings depend entirely on the particular instrument utilized and the type of microphones used. Ideally, with the right mics, you should only have to do minimal EQ'ing, if any.

Kick

There are three general types of drum sounds that engineers seem to go for: (1) the dead thud you get with one head on the drum and some type of weight (sandbag, mic stand bottom, or brick) on top of the pillows or padding in the drum, (2) the resonant ring you get with two heads on the drum and a small hole in the front head of the drum, and (3) the dull boom you get with both heads on the drum and no hole (commonly used for rap, hip hop, or techno).

The first and second type of sounds normally have a huge amount of the muddy range taken out, as much as 10 dB, in the EQ range around 100–400 Hz. They also sometimes have a high-frequency boost of a few dB around 3,000–6,000 Hz.

The third type of rap or hip hop kick often leaves in the muddy range around 100–400 Hz and a boost around 40–60 Hz for the low end. The high end, around 1,000–6,000 Hz, might actually be rolled off to get rid of the attack of the sound.

There are many other types of EQ for kick drums, but these are some of the most common settings.

Snare

The snare drum is commonly only boosted in the highs around 3,000–6,000 Hz. Sometimes a bit of low end is added around 60–150 Hz to make a thin drum sound fatter. However, the best way to get a big, fat snare sound is to use a big, fat snare drum. Occasionally, it is necessary to take out some of the muddiness around 100–400 Hz. Some snare drum sounds have a midrange "flap" or edginess around 800–1,000 Hz that needs to be taken out to smooth out the sound.

Hi-Hat

It is often necessary to take out just about all of the low end to get rid of the bleed from the kick drum. If you have a highpass filter, you can roll off the entire low end up to around 300–700 Hz. It is also quite common to roll off the muddiness in the bleed from the rest of the drum kit (around 100–400 Hz). Occasionally, it is nice to add a just a bit of super high frequencies, around 10,000 Hz, for a nice, bright sizzle up top. Also, every so often, it is necessary to take out any irritating frequencies in the midrange between 1,000 Hz and 4,000 Hz. However, if taken out too much, the hi-hat will sound too dull; a thin bandwidth is helpful in this situation.

Bass

There are two main types of bass guitar sounds: (1) the "string-y" sound that has a bright high-end, and (2) the "round" sound that has very little high-end. On some bass guitars, it is necessary to take out some of the muddiness around 100–400 Hz. However, if taken out too much, the bass will sound too thin and wimpy. For the bright stringy sound, it is also often necessary to boost the highs (much more than you would think when in solo) around 2,000–30,000 Hz. Occasionally, it is fun to boost the low end of the bass around 55 Hz to add a solid bottom.

Guitar

Most commonly, guitars only need to be brightened up around 3,000–6,000 Hz. Occasionally, it is necessary to take out some of the muddiness around 100–400 Hz.

Vocals

Vocals vary tremendously depending on the sound of the person's voice and the mic used. It is quite common to not EQ vocals while recording because it can be difficult to find the exact same EQ in future vocal overdub sessions. Unless there is a problem with the microphone, I highly recommend waiting until you have the whole mix up before EQ'ing your vocals. This is fine because vocals are normally EQ'd so little anyway.

Not only are we hypersensitive to midrange frequencies (where vocals hang out), but we are also extremely hypersensitive to the natural sound of vocals. We know what a voice should sound like better than any other sound in the world. Therefore, it is critical to be sparing with any vocal EQ.

Vocals are often boosted just a couple of decibels around 3,000–6,000 Hz. Occasionally, it is necessary to take out a bit of muddiness around 100–400 Hz and a bit of irritation around 3,000 Hz or 4,000 Hz. The irritation sometimes comes from the harmonic structure inherent in the sound, but it can also come from a cheap or bad microphone. It is often helpful to use a highpass filter to roll off all low frequencies below 60 Hz in order to get rid of any rumbly noises or bleed from any bassy instruments.

Piano

The EQ of a piano is highly dependent on its inherent sound and the style of music being played. Rock and roll has a much brighter sound than classical piano. Much less EQ is done to a solo piano piece than to a piano part placed in a mix.

Commonly, a bit of mudd is taken out around 100–400 Hz, and a bit of boost is given around 3,000–6,000 Hz. Depending on the piano, sometimes there is a muddy midrange area, around 800–1,000 Hz, that needs to be dropped ever so slightly. If the top of the piano is closed (or almost closed), it will normally create a resonate frequency in the midrange that needs to be taken out.

Organ

The EQ of an organ is very similar to that of the piano: bring down the mudd and boost the highs. With a Leslie speaker on a Hammond organ, sometimes it is necessary to roll off some low end to get rid of the rumble.

Horns

Of course, trumpets normally don't need much of anything for the high -end. Occasionally, they are mellowed out by dropping an edgy, irritating frequency to around 3,000–4,000 Hz. You can use a highpass filter to get rid of the low-frequency bleed or noise, since the horn doesn't go that low.

Sax is often only brightened a little around 3,000–6,000 Hz. Depending on the sax and the microphone, you sometimes have to take out the "honky" sound around 800–2,000 Hz.

Acoustic Guitar

It is important to take out the mudd around 100–400 Hz, more or less, depending on the mic placement (keep the mics away from the sound holes). A little boost around 3,000–6,000 Hz adds clarity, and a minor accentuation around 10,000 Hz adds sparkle.

Interesting EQ

Certainly there are those who don't use the traditions or history of EQ to set their EQ. Some people are very intuitive about their frequencies. Think about it: how did the first engineers know how to EQ sounds? One way was to make it sound natural. But what is *natural* when it comes to the sound of a piece of sheet metal? What is a *natural* EQ for a lot of the unnatural sounds we find in synthesizers?

The EQ of a sound is sometimes based on sounding interesting rather than sounding natural. Therefore, the question becomes what makes a sound interesting?

Interesting comes in various flavors. One way is to simply not make the EQ natural. Another is to EQ it so that the maximum complexity of the sound shows through. This means to use the EQ to even out any excessive peaks in a sound. Check out this spectrum analysis of a sound.

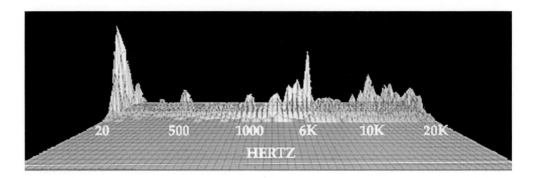

Notice the peaks around 20 Hz, 6000 Hz, and 10,000 Hz. If you were to listen to this sound, you would primarily hear these three loudest frequencies. With EQ, you can turn down the peaks, to hear more of the entire frequency spectrum. The sound appears to be more complex, more interesting. When a sound is more complex, it sustains repeated listening better, because the closer you listen to it, the more you hear. This is a common production value for many major producers: to make a sound appear as complex as possible by getting rid of the peaks.

On the other hand, there is always the counterculture (thank goodness). Instead of complexity, some have now reverted back to using simple sounds. For example, Phil Collins recorded a song with an 808 snare, the tinniest and cheapest-sounding electronic snare in the world; it sounds like "doooh." This proves that "interesting" is subjective.

But regardless of whether you EQ a sound to be interesting or natural, it is important to make sure that the EQ of the sound also works with the other sounds in the mix. As previously described, the sound should have appropriate highs, midranges, and lows relative to all of the other sounds. As with "balancing volume relationships," it is commonly most desirable to have the EQ of all of the instruments as even as possible, so that they blend well. However, it is often desirable for certain instruments in a mix to be unusually bright, dark, or midrangey. In fact, sounds can be made to sound more similar to each other or more dissimilar. A lead instrument might be made more cutting and abrasive so as to really grab attention. An instrument might be given extra bass to make the song more danceable or just to excite the listener.

The particular EQ of a sound and its relationship to the rest of the sounds in the mix creates another musical and emotional dynamic for the engineer to wield in his or her quest for the perfect mix – even though perfection has no limits.

Level 2 Dynamics: Overall Patterns of Equalization

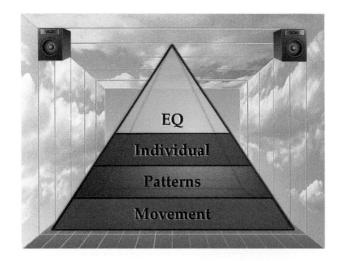

Out of all of the twelve dynamics I'm covering (three levels each for volume, panning, EQ, and effects), the overall EQ of a mix is the most important to get right. If it is not correct, everyone knows it – including babies and grandmas. And you know it, too. You can work on a mix for 10 hours, but when you listen to it the next day or so, within 5 seconds you know if the overall EQ is right or not. 5 seconds! When anyone listens to a mix, the first thing they hear is the overall EQ. This combination of all EQ settings together in the song creates a much stronger dynamic in a mix than any single individual EQ setting.

If you don't already know what overall EQ is, just change from station to station on a radio. Within the same 5 seconds, you can probably guess the style of music just from the overall brightness, midrange, and bass. Each style of music has its own typical EQ curves.

The engineer EQs each instrument so that the overall EQ of the song will sound like a particular type of music usually sounds. For example, country has an overall natural EQ. Heavy metal has more of a boost in the cutting midrange frequencies. Rap and hip hop have the low-end boost.

The type of song also can determine the general EQ of the mix. For example, you just might make the overall EQ a bit more edgy and cutting for a song about chainsaw murders, whereas a song about sweet and sensitive love might have an overall EQ that is very pleasing and conducive to mush.

Aside from the style of music and the song, often certain engineers have their own style that they prefer. These engineers tend to create mixes with overall EQs they happen to be partial to. Here are some typical – in fact, stereotypical – overall EQs for different styles of music. These EQ curves are sometimes very specific for certain styles of music.

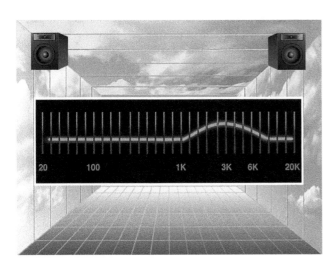

Visual 173.
General Overall EQ for
Heavy Metal

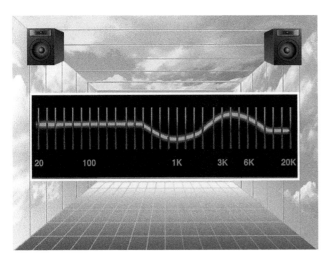

Visual 174.
General Overall EQ for
Jazz

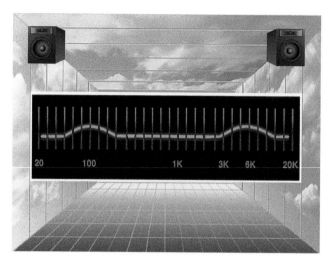

Visual 175.
General Overall EQ
for Country

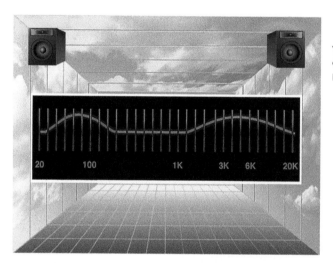

Visual 176.
General Overall EQ for
Rap, Hip Hop, Electronica

Because it is so critical to get the overall EQ right, it is often a good idea to pull out a CD of a similar style of music just before you record the final stereo mix. Compare the CD to the mix you have going and see if it is in the ballpark. And, don't be ashamed at all about going out to the car to listen. Besides, you know your car's system better than anything. Also, listen to it on as many different types of speakers as possible (including a small boom box). I used to even take final mixes to big stereo stores and pretend I wanted to check out various systems. Of course, I told them I wanted to use my own CD to compare systems (they caught on pretty quickly after a few visits, though). Whatever it takes, get it right so it matches the traditional overall EQ for that style of music regardless of what speakers you are listening to.

Level 3 Dynamics: Changing Equalization

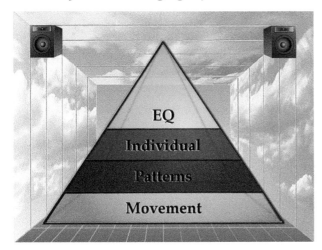

Because we are so limited as to how creative we can be with EQ, making EQ changes in a sound while it is playing creates an extremely noticeable dynamic. If not appropriate for the song, this could be quite distracting. If appropriate, it can be quite effective.

Making an EQ change at a break in the song is the most natural way to create a dynamic. Jethro Tull did it in the song "Aqualung," when the voice changes to a telephone EQ. Pink Floyd also did it in the album *Wish You Were Here*, when the acoustic guitar sounds like it is in a little box.

If you change EQ at the beginning or end of a section in a song, it is not nearly as noticeable as when you make a change in the middle of a part. Probably the most bizarre effect is to actually sweep the frequency knob of an EQ in the middle of an important part, such as a lead solo. Doing this totally takes the focus away from the music itself; however, if done in a way that works with the music and song, it can be extremely cool.

It is considered very unusual to actually change EQ settings during a mix, however it can be very interesting especially for those mixing dance, electronica, or techno.

As with everything, the musical and emotional dynamics that can be created with EQ – EQ placement, overall EQ, and changing EQ – all depend on the style of music, the song and all of its details, and the people involved.

Section C: Panning Dynamics

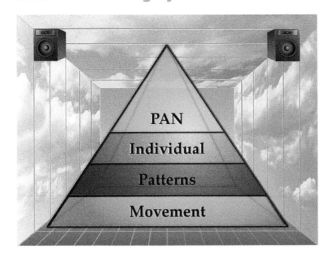

As with volume and equalization, there are three levels of dynamics that can be created with the placement of panpots on a mixer. First, a sound will be perceived differently depending on where it is placed in the mix, left to right. For some instruments, the traditions for the specific placement of left to right have become very strictly enforced. Panning is also based on the relationship of a particular instrument to the panning of the rest of the instruments in the mix. But when you create an overall pattern of panning, you establish a much more powerful musical and emotional dynamic. For example, a lopsided mix left to right will come across quite differently than a mix that is balanced from left to right. Finally, when you move a panpot during a song, you are creating a dynamic almost that is unusually powerful. Now let's explore the three levels of dynamics.

Panning Dynamics

Level 1 Dynamics: Individual Panning Settings and Placement

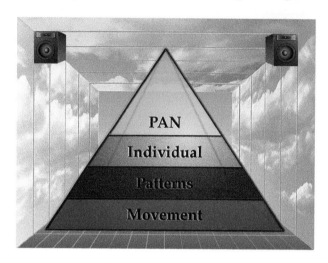

If you follow the traditions, you create a dynamic that is transparent and lets the music show through more. Whereas, if you don't follow tradition, you are then considered to be "creative." Unusual panning can actually create tension; this can be cool if appropriate. Let's go through typical panning placement for some well-known instruments and sounds.

Kick Drum

It is rare that the kick drum is ever placed anywhere except in the middle, exactly between the speakers.

Visual 177.
Kick Drum Panned to Center

You won't go to jail if you place it somewhere else, but it has become commonly accepted that it should be centered. It is interesting to speculate why the kick has been relegated to the center. First, the kick often takes up a large amount of space in the world of imaging. There is simply more room in the middle. Also, the kick drum has so much energy it commands our attention. We are often compelled to turn and face the music, especially loud and powerful sounds. Therefore, if you are facing the kick in the middle, your peripheral vision (or hearing) can see the other sounds better. If you were facing a kick on one side, then you would be off balance.

Aside from the imaging, there is another reason, based on physical reality, for the kick to be placed center: when a sound is in the middle, you have two speakers carrying the sound instead of one. The speakers don't have to work as hard, especially with big sounds like kick drums and bass guitars. Therefore, technically it sounds better when in the middle. Also, the kick drum is normally in the center of most drum kits.

Regardless of the reasons for placing the kick drum in the center, it has become a pretty strict tradition. If you place it anywhere else, watch out . . . you could be admonished for being too creative.

Two kick drums, or a double kick, present an interesting dilemma when it comes to panning. The main deciding factor depends on how often the second kick drum is played. Some people will pan them slightly left and right, others will place the main kick in the center and only pan the second kick slightly. To pan the two kick drums completely left and right is highly unusual, or creative, but has been done. I've even panned the two kicks left and right only at the moment of a classic double kick roll.

Snare Drum

The snare drum is often placed in the middle.

Visual 178.
Snare Drum Panned to Center

A large number of engineers do place it a bit off to the side (even more so in jazz) because the snare drum is off to one side in a real drum set. If the snare drum sound is huge with a bigger sound (large snare, played hard, and more reverb), it is more commonly placed in the center, for the same reasons that a kick drum is placed in the center. It is interesting that the snare has come to be so commonly placed in the center when it is actually so far to one side on most drum sets.

Hi-Hat

The hi-hat is often placed about halfway between one side and the middle. (I'll discuss which side later.)

Visual 179.
Hi-Hat Panned Halfway Between
Left Side and Middle

This is also interesting because the hi-hat is normally all the way to the side on a real drum kit. When the mix is busier, the hi-hat might actually be panned all the way to one side. This is also the case when the mix being created is meant to be "spatial." Meanwhile, in house music and hip hop, not only can the hi-hat be panned anywhere, it is sometimes moved during the mix and is occasionally panned far left with a delay panned to the far right.

Tom Toms

For the most part, toms are panned three different ways. First, and most commonly, toms are panned from left to right (or right to left), but the stereo spread is pulled in a little like this:

Visual 180.
TOMS L/R – Not So Wide Panning

This is commonly done when there are fewer toms (two or three). Second, in order to provide maximum fun, tom toms are commonly spread completely left to right (or right to left), especially when there is a large number of toms.

Visual 181.
Toms Panned Completely Left
to Right

However, for natural panning, the toms are sometimes placed between the speakers exactly as they are on the actual drum set.

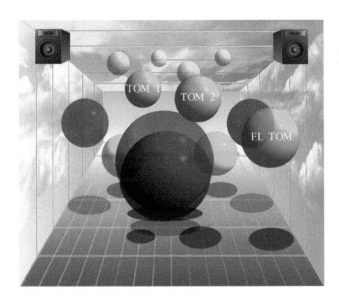

Visual 182.
Toms Panned Same
as on Drum Kit

A floor tom is normally placed on the far side. However, occasionally the floor tom will be placed in the center for the same reason we normally put a kick drum in the middle – because it is so powerful, commands so much attention, and will sound better when both speakers are carrying the sound.

The discussion of tom placement brings up an interesting question: should the toms be panned from left to right, as if from the drummer's perspective . . .

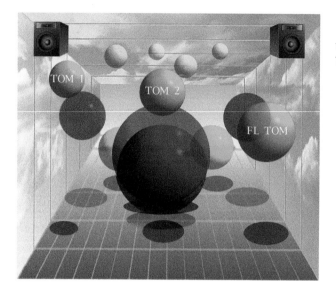

Visual 183.
Toms Panned Left to Right

. . . or from right to left, as if from the audience's perspective?

Visual 184.
Toms Panned Right to Left

Those who do live sound wouldn't be caught dead with the toms panned left to right because they always see it this way in a live show. It seems that just as many engineers pan from left to right (just like we read). If the band is being recorded live, or if the band is being recorded as if they were live, then the toms should probably be panned right to left, from the audience's perspective, because there is an audience. Even though it really doesn't matter which way you pan the drums in a mix, most people seem to have very strong feelings about the matter; so it's good to be aware of the preferences of the people you are working with. Besides, it would be boring if all toms were panned the same way.

The only important point here is to be sure to pan your overhead and hi-hat microphones the same as you pan the toms. If the toms are going left to right, make sure they are going the same direction in the overheads and that the hi-hat is panned to the left – or vice versa respectively.

Overheads

Overheads are normally recorded in stereo on two tracks, and then spread completely left and right between the speakers. This allows for the maximum separation between cymbals and for the widest spread of stereo imaging. Of course, the type of imaging you get from the overheads depends on the placement of the mics themselves.

If you place the mics next to each other in the middle using the "X" technique, the imaging is often a little clearer because there is no phase cancellation. Even if you pan the mics completely left to right, you will not hear a very wide spread between the speakers.

Visual 185.
Overhead Panning When Using "X" Technique

However, some people prefer a spread that is not so wide because it makes the cymbals and drum kit sound more cohesive.

On the other hand, if you place the mics as far apart as possible, even though you do have a greater chance of getting phase cancellation, you will get a wider spread of cymbals between the speakers. As with toms, this might be preferable if you have a large number of cymbals or if you have a busy mix with a lot of sounds.

Visual 186.
Overheads Panned Completely Left to Right

The closer the mics are placed to the cymbals, the clearer and more up front the image will be between the speakers (put a mic closer to anything and it will sound more present), but the cymbals will also sound more edgy, which could be fine for rock and roll or punk.

Drums as a Whole

It is interesting to note the way that drums have been panned throughout the history of recording. The Beatles placed the vocals in one speaker and the rest of the band in another. Though this was, in reality, a mistake. They meant for the two tracks to be mixed down to mono when the record was made, but the mastering engineer decided to be creative. Many traditional jazz groups have placed the entire drum set in one speaker. The obvious advantage of doing this is that it leaves a huge amount of space between the speakers for the rest of the band. The big disadvantage is that the separation between individual pieces of the drum set becomes obscured.

Visual 187.
Mix with Drums Panned to
One Side

In order to obtain the most natural panning of a drum set, try this: pan the overhead mics on the drum kit completely left and right, listen to where each drum seems to be between the speakers, and then pan the mic of each individual drum exactly where you hear it in the overhead mix. This will give you the clearest imaging you can obtain because the image of the instrument in the overhead mix is in the exact same place as the image in its own mic. If they are not panned the same, each sound is actually spread in stereo between where it is in the overheads and where its own mic is placed, making the image less precise.

This technique is good when you want to create a mix that has the clearest imaging. However, most engineers will often pan the toms wider anyway – for maximum fun.

Bass Guitar

The bass guitar is most commonly placed in the middle because it is so large and commands so much attention, like the kick drum.

Jazz and similar types of music will sometimes place the bass off to one side. Aside from the style of music dictating the panning, a bass is normally only panned off to one side if the part being played is a lead part. When this is the case, the bass is often a much brighter and cutting type of sound. When the bass sound is thinner, there tends to be more room for the bass to be placed off to one side.

Visual 188.
Mix with Bass Guitar Panned to Center

Lead Vocals

It is almost against the law to place a lead vocal anywhere except smack dab in the middle. Placing a lead vocal in the center has become such a strong tradition that it actually creates an element of tension when panned anywhere else. Therefore, it should only be done when it is appropriate for the song. For example, it might fit a song about unbalanced psychotic behavior, or a song about feeling "out here all alone, on my own."

If a vocal is recorded in stereo with two mics, doubletracked, or made into stereo with a time-based effect, the two sounds are normally spread evenly left to right. Sometimes they are placed at 11:00 and 1:00.

Visual 189.
Lead Vocals Panned at 11:00 and 1:00

Sometimes they are placed at 10:00 and 2:00. But occasionally, they are placed completely left and right.

Visual 190.
Vocals Panned Completely
Left and Right

Background Vocals

The panning of background vocals often depends on the vocal arrangement. When there is only one background vocal, if you pan it to the center, it gets in the way of the lead vocal. You could put it off to one side or the other, but this makes the mix unbalanced. Commonly, a single harmony is made into stereo with two mics, doubletracking, or a time-based effect (delay, chorus, flanging, etc.). Then it can be panned in stereo, creating a balanced mix from left to right.

Visual 191.
Background Vocal Made into Stereo
with Fattening

If there are two background vocals singing the same part in unison and you place the background vocals completely left and right, they will "pull together," creating a line of vocals.

Visual 192.
Two Background Vocals
Pulling Together

If the background vocals consist of different harmony parts, they won't pull together as much. The more different the sound of the voices that are doing the harmonies, the less they will blend together, the more they will stay separate. If the same person does all of the parts, the more they will pull together.

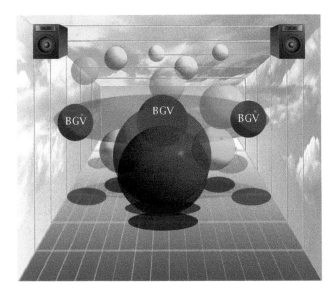

Visual 193.
Three Background Vocals
Panned Separately

Background vocals are commonly recorded in stereo, doubletracked, or made into stereo with a time-based effect, and then spread left to right. As you can see, there is a wide range of possibilities.

Visual 194.
Seven Background Vocals Panned to Seven Different Places Combined with Variety of Fattening

The style of music can also make a difference in panning. For example, in country music, many engineers will only pan the background vocals from 11:00 to 1:00 or from 10:00 to 2:00, in order to give the impression of a tight-knit harmony group.

Piano

A solo piano is almost always panned completely left and right in stereo. The bass strings are panned to the left and the high strings are panned to the right, because this is the way a keyboard is laid out. It is funny that this is probably the strictest rule of all when it comes to panning. You better shoot the piano player before you pan the high end to the left. Of course, as you might guess, this rule would hardly apply in hip hop nor, especially, when mixing dance music.

When in a mix, a piano is still commonly panned in complete stereo, just because it sounds so nice that way. That is, if there is room in the mix. However in certain styles of music, such as country, even in a busy mix, the piano is still commonly panned completely left to right. Sometimes it is pulled in a little bit or panned more to one side to leave room for other sounds. In some busy songs, the piano is panned in one spot to leave room for the rest of the mix.

The panning also depends on the type of musical part being played. If the part is full of rich sustaining chords, it will tend to be panned in full stereo. On the other hand, if it is very staccato and rhythmical with many single notes, it might be panned in one single spot.

One argument for panning the drums from the drummer's perspective is that if you were to place the hi-hat on the right, it would be hidden by the high end of the piano. There is less masking from the high end of the piano when the hi-hat is placed on the left.

If you don't have enough tracks to record a lead piano in stereo, you could actually pan the piano from left to right when the pianist plays from low to high notes. When they go up the keyboard, pan to the right; when they play lower notes, pan to the left. Cheap, fake stereo, but it works.

Visual 195.
Panning with High End of Piano on
Right and Hi-Hat on Left

Guitars

Panning guitars is based on concerns similar to those for piano and keyboards. Often the guitar is placed wherever there is room for it, based on the placement of everything else. If the part is more staccato and rhythmical, it might be panned to one spot. If it is more legato with more chords, it might be spread in stereo with fattening. You might also do this if you want the guitar to be more interesting or present.

Horns/Strings

It is interesting that horns and strings are almost always spread in stereo completely left and right across the stereo field. As with toms, the more horns, the wider the stereo spread. The horns or strings can be recorded with more than one mic, played twice, or a time-based effect can be used to make the instruments stereo. The horns or strings might not be spread completely in stereo (partial stereo or mono) if there isn't enough room in the mix.

Effects

Effects like delay, flanging, chorusing, phasing, harmonizer, and reverb can be panned separately from the instrument sound they came from.

Delay

When the delay time is greater than around 30 ms, it is perceived as a separate sound. This separate "sphere" is often placed wherever there is room for it (crowd control). The further from the original dry sound that the delay is panned, the more intense the dynamic created. However, it is easy for this effect to overwhelm the song. Sometimes it is quite effective to pan the delay right on top of the original dry signal.

When the delay time is less than 30 ms, the sound is stretched between the speakers. As previously covered, this effect is called fattening. The primary consideration for using fattening is the duration of the dry sound. Fattening is rarely used on staccato sounds. They just seem to take up too much room for something so quick. If you were to use fattening on a sound with a short duration, you might not pan it so wide – maybe only 11:00–1:00 (or just around the dry sound itself if it is panned off to one side). On the other hand, sounds that are longer in duration (legato) are often panned completely left and right in stereo with fattening.

Once the drums and bass are pulled up and mixed, you should calculate which of the rest of the rhythm instruments will be spread in stereo with fattening and which ones will be in one spot.

Often the style of music dictates this; however, you normally have a lot of choices, based on the number of instruments/sounds in the mix, and whether each sound is staccato or legato.

As an example (with limited equipment), a common layout is to place a guitar on the left and spread it with some fattening panned completely left and right. Then you bring up just enough fattening so that you get a good balance of clarity and fullness – a clear guitar on the left and full stereo fattening. Then you place a keyboard on the right and send it to the same stereo fattening – again, just the right balance of clarity of the original sound coupled with fullness of the stereo spread. Do the same with two tracks of background vocals panned completely left and right – send each of them to the stereo fattening. Then, bring up the lead guitar (or other lead instrument) in the middle and send it to the same fattening so it is spread completely left and right. You could go ahead and send a large amount so that the guitar is nicely spread. Then, finally, place the vocals in the center and send them to the same fattening. It would actually be preferable to send them to some fattening that is not panned so far left and right – more like 10:00 and 2:00. Only send them to the fattening enough so you can just hear the effect (you normally don't want to make the vocals sound too unnatural). The whole layout might look something like this before adding the fattening (less the drums and bass):

Visual 196.
Simple Mix without Stereo Fattening

And it might look like this with all of the stereo fattening described above:

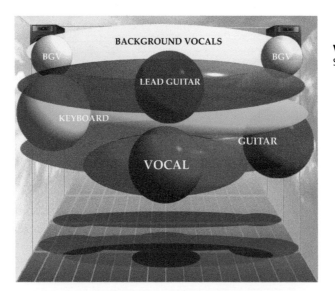

Visual 197.
Simple Mix with Stereo Fattening

What a difference, huh?

Flanging, Chorusing, and Phasing

Because these effects are based on a short delay time, like fattening, they are panned based on the same criteria. The only difference is that these effects are much more noticeable; therefore, they might not be panned quite as wide as fattening.

Reverb

Reverb is most commonly placed in stereo, completely left and right. This is to simulate the natural sound of reverb in a room where it comes from everywhere around you. This is especially common when putting reverb on drums.

Sometimes, though, reverb is panned in just one spot. For example, you could put a guitar in the left speaker and place the reverb in the right speaker. It is also quite effective to place the reverb right on top of the dry sound. For example, place a keyboard on the right and put the reverb in the right speaker also. This is especially effective when using short reverb sounds or gated reverb. This tends to make the sound appear to be in its very own space, which creates a unique spatiality for the mix. This also preserves your limited space in the mix for other things. Obviously, when reverb is panned to only one spot, it takes up way less space than when panned completely left and right.

Any placement of individual instruments other than in the above norms might be considered creative or unusual, depending on your perspective. Again, the key is to make sure they fit the music and song.

Level 2 Dynamics: Patterns of Panning Placement

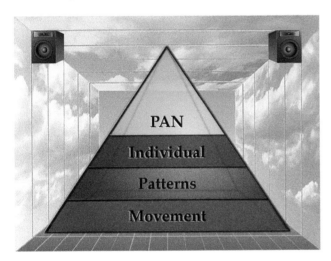

The overall pattern created by all of the panning settings together is much more important to the style of mix than any individual panning. The panning of all the sounds together in a mix can create a huge number of different structures of mixes. The type of music tends to make the biggest difference as to the overall panning in the mix. And again, certain styles of music have developed more strict traditions than others. Also, the details of the song, especially the density of the arrangement, can affect the overall placement left to right. Often the more instruments in a mix, the wider the overall panning.

There are a few different ways to look at the types of patterns that can be created. Overall panning is based on natural panning, balanced versus lopsided panning, and crowd control.

> Before stereo became popular, mixes had to be created so that all the sounds could be heard with no panning at all. This is good to keep in mind. In fact, you should always double-check your mix in mono to make sure that it is OK in the first place, before using panning to create clarity. After all, listeners are often not seated in the correct position between the speakers to hear true stereo.

Natural Panning

Panning is sometimes spread so that it corresponds to the placement of the band, as if on stage, or the way they are set up in the studio.

Visual 198.
Panning as if Onstage

Normally drums are not panned as they are on a live stage (they would all be panned to the center anyway), except for big band music. However, drums might be panned exactly the way the drum set is physically set up in the studio.

Visual 199.
Natural Panning of Drum Kit

Again, it is often the style of music that determines if panning is to be natural. For example, you can practically do whatever you want in hip hop or techno; whereas in big band music, it is very important to pan everything in the same way that the band sets up onstage (the horns, that is; drums and piano are often still panned completely left and right). Acoustic jazz is also sometimes panned just the way the band sets up live.

An engineer will sometimes place the musicians in the studio as if they were live onstage, just so they feel comfortable. For example, a folk group or chamber orchestra is commonly set up in a semi-circle out in the studio, then panned exactly the same way in the mix. By preserving the panning relationships in the mix, you keep the phase relationships consistent. Therefore, the imaging will be better. This helps to create the illusion that you are there with the band.

In classical music, it is an extremely strict rule that the panning is done exactly the way the orchestra sets up (otherwise you will go to jail). In fact, there are very particular rules when it comes to setting up an orchestra onstage so that everyone can hear the rest of the orchestra correctly.

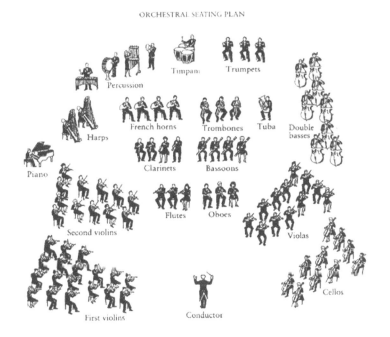

Visual 200.
Orchestra Seating
Arrangement

Here is what a mix would look like with panning that is the same as the seating. (Remember that up/down placement is a function of frequency, or pitch, of the instrument).

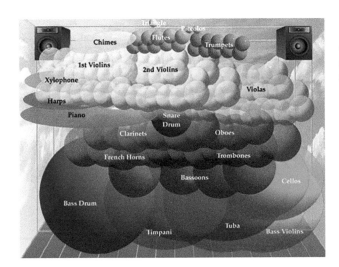

Visual 201.
Orchestral Mix with Panning,
Based on Seating

This visual shows all the sounds in the orchestra at once. Normally all sounds aren't playing all the time (especially for the low-frequency instruments).

Preserving this relationship does create a more realistic view of the orchestra in a mix. However, doesn't it just make you want to break the laws? In fact, it might make sense to balance out the mix more like this:

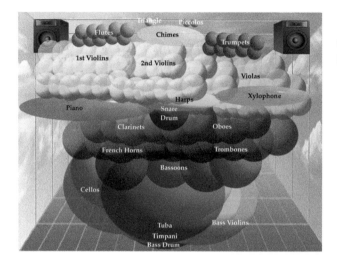

Visual 202.
Balanced Orchestral Mix

Balanced Versus Lopsided Panning

Natural panning can be . . . natural, but these days it is actually quite acceptable to not pan instruments as they are on stage. This pallet between the speakers is a different medium than that of a live show. Why not utilize it to its fullest? You can create a fairly intense dynamic by making the mix either symmetrical or asymmetrical. A symmetrical (balanced) mix looks like this:

Visual 203.
Symmetrical (Balanced) Mix

An asymmetrical (lopsided) mix looks like this:

Visual 204.
Asymmetrical (Lopsided) Mix

A symmetrical mix might be used to create a sweet emotional dynamic appropriate for a balanced type of song, such as a love song, a ballad, or a song about finding balance in life or a relationship. Whereas an asymmetrical mix might be used to create a bit of tension appropriate for an unbalanced type of song, such as a song about psychotic, unbalanced behavior.

A mix is often made to be balanced or lopsided at each frequency range. For example, if you are creating a symmetrical mix, you might put a hi-hat on the left and place a shaker or acoustic guitar on the other side to balance the high-frequency range. In the midrange, you might put a guitar on the left to balance a midrange keyboard on the right. In the bass range, the kick and bass would be placed in the center.

Visual 205.
Mix Balanced at Each Frequency
Range

On the other hand, if you're creating a lopsided mix, you might put all of the high-frequency sounds on one side and put the midrange instruments on the other side. Then for a bizarre effect, put the bass guitar on one side and the kick drum on the other side.

Odd panning structures inherently strike an odd chord in us. Just because this type of mix is out of the ordinary, an additional component of tension is created.

Visual 206.
Unbalanced Mix at Each Frequency Range

As you can see, the possibilities are endless, depending on the song and what you want to do. Creating balanced versus lopsided mixes can be an especially effective dynamic when it is appropriate.

Crowd Control

Probably the most basic and common consideration has to with placing sounds separate from each other . . . wherever there happens to be space in the mix.

When you realize that the space between the speakers is a limited space, and when more than one sound is in the same place they mask each other, then panning mostly becomes a matter of crowd control. How close are the members of the crowd to each other? Are they far apart, just touching, overlapping, or right on top of each other?

We often pan things as far from each other as possible, so each can be heard clearly. Such a dynamic might be appropriate for certain types of music, like acoustic jazz, folk music, bluegrass, and even hip hop and rap.

Visual 207.
Clarity Style of Mix:
Acoustic Jazz

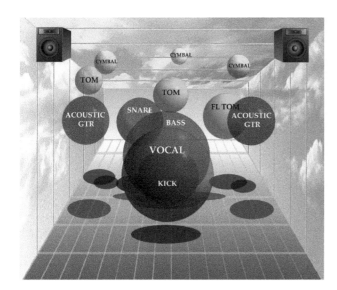

Visual 208.
Clarity Style of Mix:
Folk Music

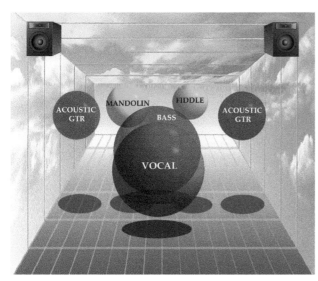

Visual 209.
Clarity Style of Mix:
Bluegrass

On the other hand, sounds may be panned to overlap in order to create a wall of sound, making the mix seem more cohesive. This is commonly done in heavy metal, alternative rock, new age, and some dance music.

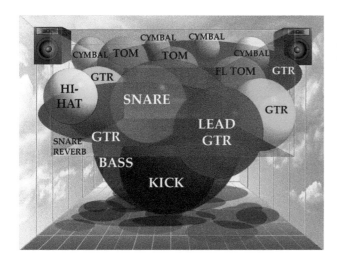

Visual 210.
Wall of Sound Style of Mix:
Heavy Metal

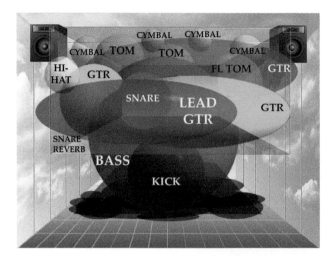

Visual 211.
Wall of Sound Style of Mix:
Alternative Rock

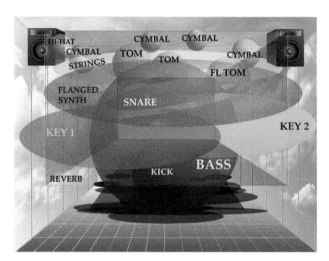

Visual 212.
Wall of Sound Style of Mix:
New Age

Besides the way that sounds overlap or not, there is also the difference between panning an entire mix as wide as possible between the speakers . . .

Visual 213.
Mix with Extremely Wide Panning
Overall

. . . versus not so wide.

Visual 214.
Mix with Panning Not So Wide
Overall

   The advantage of panning things as wide as possible is that it provides more space between the sounds, thus creating more clarity. The disadvantage is that it can make the band sound less cohesive. By making the spread narrower, the band and all of its parts sound more like they are playing together. It all depends on the density of the mix and the style of music.

   As I'll discuss later, with 3D sound processors and surround sound, you actually have more space to work with. Therefore, the possibilities for panning and placement are expanded tremendously.

Level 3 Dynamics: Changing Panning

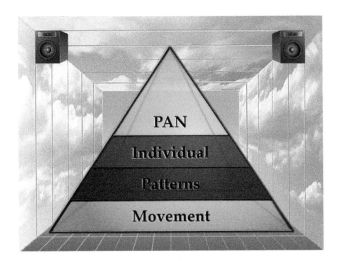

Movement of a sound from left to right during a mix creates such an intense dynamic that most engineers reserve such dramatic creativity for special occasions.

   There are a number of ways that sound can be moved from left to right, creating innumerable patterns of movement. First, you can move the sound back and forth in varying degrees. Possibilities range from short, minuscule moves to wide, sweeping moves that span the entire distance from speaker to speaker.

Visual 215.
Wide Versus Narrow
Sweeping Panning

You can also pan sounds at different speeds, ranging from pans that move slowly to pans that zoom back and forth between the speakers. You might cause some serious goose bumps by making the speed, or rate, of the pan equal to, a fraction of, or a multiple of the tempo in the song. And aren't goose bumps the goal most of the time?

Changing panning is so intense that it will usually pull attention away from the song itself. However, if the panning is done skillfully, it contributes to the music as if the panpot is an instrument itself. Obviously, when it is appropriate for the song, this can be a great effect. Hendrix did it a lot, especially in the song "Crosstown Traffic." Led Zeppelin went bananas in "Whole Lotta Love." Extensive movement of panpots has become quite common in all forms of dance music: electronica, techno, house, and trance.

One of the most intense, fun, or chaotic things you can do is move the panning of multiple instruments in various ways all at once. Such a strong dynamic is normally reserved for songs whose essence is already wild and crazy. Also, hopefully, the band will let you.

As you can see, you can create a plethora of emotions by how you set or move panpots in a mix. If you set them based on tradition, the dynamic is often unnoticeable. However, if you set them different from the norm, you create an emotional dynamic. If you set all the panpots to create an overall pattern, you are really wielding some power. A balanced mix will probably fit in and not be noticed. But if you make a lopsided mix, it will more than likely stick out, almost as if it is another musical part in the song. Finally, if you move the panpots during the mix, you aren't fooling around. Go for it, if appropriate.

Section D: Dynamics Created With Time-Based Effects

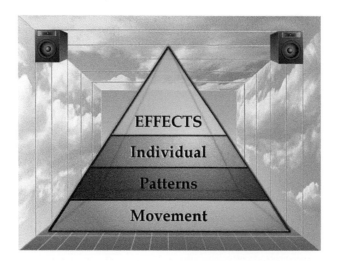

There is an incredibly wide range of effects, and the dynamics that they create range from subtle to shocking, mesmerizing, and world-changing. Furthermore, when you use multiple effects together to create an overall pattern, you can elicit a conglomeration of feelings and emotions that can be overwhelming or just good, clean fun. But changing the levels or parameters of effects during the song opens up entirely new avenues of creative expression. I'll now go into more detail on each of the three levels of dynamics that can be created with effects.

Level 1 Dynamics: Individual Effects Placement and Relative Settings

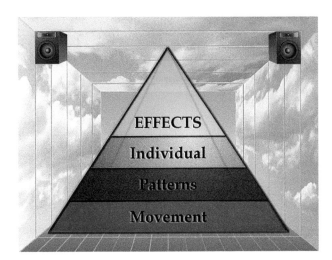

Each and every effect in the studio has its own world of emotional dynamics associated with it. For example, reverb creates a more spacious (pick your own adjective) feeling:

Visual 216.
Spacious Reverb

Long delay times create a dreamy effect:

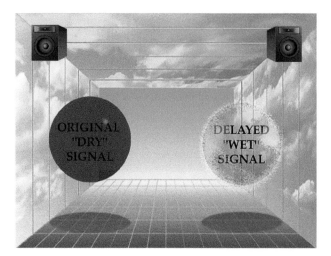

Visual 217.
Dreamy Delay

And flanging brings up a floating, underwater-type feeling.

Visual 218.
Underwater Flanging

When you change the parameters of the effect, you also change the feeling that it creates ever so subtly. To learn the intricacies of the feelings that different effects create, simply play with them. As with any instrument, practice makes perfect. Get to know your tools. Get to know them so well that you can then create art with them.

Regardless of the type of feeling that an effect adds to a mix, time-based effects, such as delay, flanging, chorusing, phasing, and reverb, all add more sounds to the mix, filling out the space between the speakers. Therefore, they all add a dynamic of fullness to a mix. The question to ask yourself is whether more fullness is appropriate for the style of music or song.

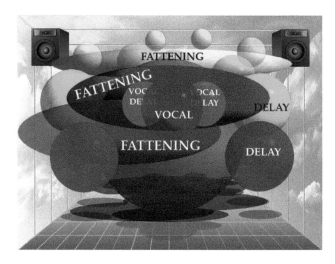

Visual 219.
Mix with Lots of Different
Delays Filling Out the Mix

As previously covered, fattening stretches a sound between the speakers, filling out the mix.

Visual 220.
Fattening

Flanging, chorusing, and phasing are also based on short delay times, so they, too, will tend to make the mix sound bigger and fuller.

Visual 221.
Flanging

And, of course, reverb is really made up of hundreds of delays, so it takes up a lot of space in a mix and fills out the mix tremendously.

Visual 222.
Reverb

All effects make a mix fuller, bigger, and badder (depending on your perspective). However, they also make a mix busier, so watch out.

Level 2 Dynamics: Patterns in Effects Placement

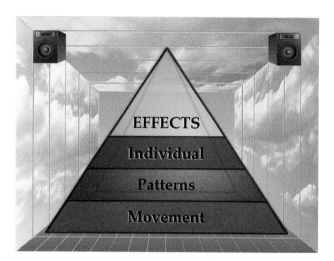

Sparse Mixes

Sometimes you add very few effects to keep the mix clear and sparse, with plenty of space between each sound. For many styles of music, like folk, bluegrass, and many forms of jazz, it is required that there are virtually no effects to obscure the pure clarity of the natural sounds. Also, you shouldn't obscure the natural beauty of a pure song, if that is what you have. After all, it is the song that counts.

Full Mixes

Time-based effects add extra sounds to the mix. When you add a delay, you now have two sounds. Add feedback and you could have ten sounds. Add reverb and you have added hundreds of sounds (delays). Therefore, when you add multiple effects, you can very easily and quickly fill out this limited space between the speakers.

Sometimes it is appropriate to use effects to make a mix sound fuller and bigger, like a wall of sound. Many styles of music, such as new age, alternative rock, and heavy metal, often have a large number of effects to fill in all the spaces between the sounds, creating a full mix. In some songs, the entire song is about effects. This is often the case with rap, hip hop, electronica, techno, and space rock. In this case, it is fine to have a ton of effects.

The main thing that makes a mix sparse or full is actually the number of sounds and notes in the song. Some songs have a busy arrangement in the first place; while others have a sparse arrangement. Therefore, when you approach a mix, one of the first steps is to check out the density of the arrangement. If there is a lot going on in the arrangement, you normally use fewer effects, simply because there is not enough room left between the speakers. This is commonly the case with salsa and symphonies where so much is already going on.

Visual 223.
Extremely Busy Mix with
No Effects

However, there are times when you might want to make a busy arrangement even bigger. Often bigger is better with new age, alternative rock, heavy metal, and other hard rock. The more powerful, the more awesome. Therefore, you might consider adding effects to such a mix even if things are already crowded. Forty-eight guitars may not seem so clean and clear, but it creates such a massive wall of sound that it can blow people's minds.

It was Phil Spector who was first known for creating this style of mix. In fact, he even did it in mono. For the longest time, everybody was addicted to clarity. Then Phil started adding more and more instruments to the mix and started using reverb to really fill out the space between the speakers. His mixes were dubbed "The Wall of Sound." These days we have taken this concept to the extreme. You might take a moment to think of the songs you know that have a busy arrangement with a full mix.

Visual 224.
Extremely Busy Mix with
Lots of Effects

On the other hand, if the arrangement is sparse in the first place, and if the tempo is slower, you have plenty of room for effects. You could use fattening to fill out the space between the speakers. This just might mean the difference between a garage band sound and something that sounds like a real CD.

Visual 225.
Extremely Sparse Mix with
Fattening and Reverb

Occasionally, an extremely sparse arrangement is left that way. In this style of mix, every single sound is completely separate from every other sound in the mix. This makes each sound easily distinguishable from all the other sounds. As mentioned, bluegrass, acoustic jazz, and some folk music are commonly mixed this way. Steely Dan is a good example of this style of mixing: very few effects are used to create as clean and clear of a mix as possible.

A full or sparse mix might be appropriate depending on the type of person you are working with, the nature of the song, and of course, the style of music. However, it is important to first determine in advance which type of mix you are going for. Then ask the band what they think. Otherwise, you can waste a good amount of time going down the wrong road. In the past I've had many situations where I've collected many cool ideas (even written them down on the track sheet) while a band is recording all their parts. Then when we got to the mix, I immediately began creating an extremely cool mix, utilizing all the ideas. But before long, the band would say, "By the way . . . we don't want any effects. We like dry mixes." Now, I always ask first.

Visual 226.
Extremely Sparse Mix with
No Fattening and Reverb

Level 3 Dynamics: Changing Effects (Types, Levels, and Parameters)

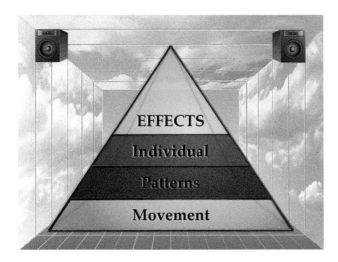

Changing the levels of effects or the parameters of effects during the mix is one of the most intense dynamics you can create. It is such a strong dynamic, it must certainly be appropriate in every way.

There are a number of ways that you can create movement with effects. You can turn up a delay on the end of a word, word, word or add reverb on the end of a solooooo. You could add flanging to a line in a song about "floating through life." Or take all the reverb off a particular line to make it more up front and personal. One of the coolest effects possible is to take an effect off of something. The absence of an effect that we have grown accustomed leaves a profound place to highlight any special instrument or lyric.

Adding or taking away an effect, or just changing the level of an effect, is such a strong dynamic that it often becomes the sole focus of attention for the moment. Therefore, the effect should be done skillfully, so that it fits in with the song and the music. Some performers, like Frank Zappa, Mr. Bungle, and even Pink Floyd, have created songs based around changing effects.

You can simply turn the amount of effects in the mix up or down, but you can also change the type of effect or change the parameters within each effect. The capabilities of doing this are much simpler now with MIDI controllable effects. You could actually have the effects change based on the pitch of the note being played. Or you could use some sort of MIDI controller to play the effects like an instrument.

Additionally, using automation with Digital Audio Workstation plug-ins, you can program parameters of effects to change in real time. And, with a time grid mode setting, you can program the changes in the effect parameters to happen in time to the music. Some programs also allow you to sweep parameters of various effects on a variety of instruments in real time.

Of course, you can only create such overwhelming dynamics if the band will let you. You might keep a lookout for those bands that write songs with changing effects in mind. This is why dance music can be so much fun to mix.

Even if a song doesn't have completely different sections where you can use completely different effects, you can still create subtler variations between the sections of a song. You might add a little more reverb to the snare for the chorus section of the song, change the type of reverb on the snare for a lead break, or add fattening on the lead vocal during the chorus. Commonly the reverb on the snare is boosted ever so slightly at the end of a song when it gets rocking.

It is extremely cool to create subtle differences in various sections of a song, so when someone is listening closely they will hear more detail. When they listen over and over and over, they will always hear something new and will never get bored.

Although technically it is not effects, it should be mentioned that in most DAW's you can also automate every parameter of a Sound Module. This opens a huge area of changing possibilities since synthesizers have so many knobs.

You now know what to listen for and you have a framework within which to remember what others are doing. Not only can you then draw from (nice word for steal) everyone else's mixing tricks, you can begin to develop your own.

Homework for the Rest of Your Life – Due (Do) Every Day

Now that I have covered all of the dynamics that can be created with all of the studio equipment available, your homework is a little more complex.

When you have the time to listen closely, check out every single sound in the mix and see where it has been placed as far as volume, panning, EQ, and effects. Then for each sound ask yourself the first question:

> Why might they have placed the volume, panning, EQ, and effects where they did?

Again, this helps you to learn all of the considerations that affect what you do in a mix with each of the four tools. More importantly, it gives you *valid reasons* that you can use to explain yourself to someone who is telling you to make the mix suck.

Then ask yourself the second question:

> Do I like the volume, panning, EQ, and effects for each sound there?

As discussed, the first time you hear a sound in a mix, you may not have an opinion. But the next time you hear it, note how it is different. Once you hear the sound a third time, if you are listening closely, you will probably be able to tell which of the three sounds seems the best to you. After you have done this for a few years, you will gain an incredible perspective on what others are doing.

Most importantly, you will develop your *own* values. Then when you go into the studio, you not only know what everyone else is doing, you know what you like in relation to the rest of the world. You will have developed your own style. Then you are a professional engineer.

And, the best part of it all, is that other people will see that you know. Ultimately, you will gain their respect. Then, when they tell you to make a mix suck, all you have to do is slowly turn around and look them in the eye. Often you don't have to say anything . . . they will know.

By the way, all professional engineers are doing this homework exercise all the time.

Styles of Mixes

I have now covered each of the three levels of dynamics that can be created with the four tools in the studio – volume, EQ, panning, and effects. Each of the four technical tools creates the dynamic patterns that evoke specific emotions and feelings. However, when you combine effects, you can create a whole other level of emotions and feelings as the effects interact with each other. For example, if you want to make something really present, there are many effects that create presence, which when used together can make a sound extremely present.

Also, when you use the four tools together to create a certain style of mix, you can create a really powerful dynamic. The most powerful dynamic of all is when the style of mix is abruptly changed to a different style of mix without warning.

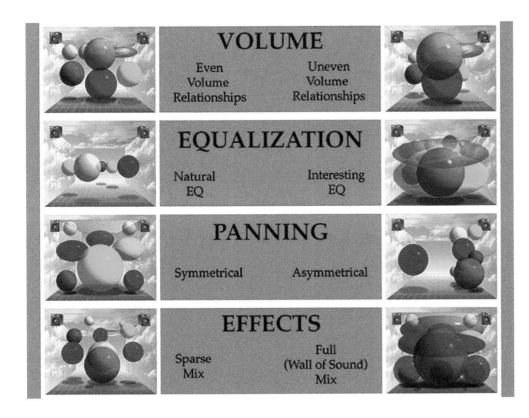

Visual 227.
Dynamic Ranges Using
Four Mixing Tools

Creating Maximum Presence

Each of the four tools can be used to make a sound appear more present; however, when you use all of them together, it will make the sound extremely present. You can make an instrument more present and out front by using volume. Compression will make it more stable so that it seems even more present. If you brighten it a bit with EQ, it will be clearer, and keeping it clean with no effects makes it more present. Panning it to the center will also help. And if it's spread in stereo with fattening, it will really seem like it's in your face. Also using 3D sound processor can bring it even closer. Using all of these techniques simultaneously will make the sound jump right out of the speakers into your lap.

Visual 228.
Clean and Clear Mix

When a mix first begins, regardless of how much depth has been created, your mind will always place the first sounds heard on a front-to-back plane around Level 2. Therefore, if you start with really low volume sounds, or sounds with a lot of reverb on them, then they will appear closer to you than they really are. The cool thing is that when you then add sounds that are much louder, or have the reverb taken off, they will seem to be excessively out front and present. This is a great illusion to use to make sounds in a mix appear more out front and present than you can imagine.

Also, to make a sound come even further out front in a mix, start with a mix that has a good amount of reverb in it. Then take the reverb off, and the sound will seem to move forward. This can make a sound jump so far out front it seems like it could kiss you in the face.

If you want to put a sound in the background, do just the opposite: turn it down, make it dull with EQ, pan it to one side, and add long delays and reverb.

Visual 229.
Distant Mix

Creating a Full Mix

By using all of the tools together to create combinations of dynamics, you can create all the different styles of mixes in the world. One style is the wall of sound. Just using multiple effects will fill in the space between the speakers quite well; spreading sounds in stereo, with delays or adding reverb, will quickly fill in every hole in the mix. But when you also boost the low end a little on each of the sounds, they will all take up more space. You can also use panning to spread sounds in stereo, if the two sounds are similar (like two mics on a guitar amp). Then when you pan the two sounds completely left and right, they will pull together so that the sound is stretched between the speakers – making the mix much fuller. Panning sounds so that they overlap a bit makes a stronger wall of sound. If you have very many sounds in the mix, making the volumes more even will contribute to a fuller mix.

Visual 230.
Making a Mix Fuller

Because the main thing that makes a mix full in the first place is the number of sounds and notes in the song, you can help make the mix fuller by having the band play more parts. For example, you might have the band doubletrack or play the same part three times. Multiple mics on one sound can also add to the density of the mix.

As you can see, using all of these techniques together can build a huge mix. On the other hand, if you want a mix to be sparser, cleaner, and clearer, do the opposite: fewer sounds, fewer effects, brighter EQ, wider panning, and appropriate volume settings.

Classic Styles of Mixes

Each of these tools can also be used to create mellow or intense dynamics. When you use them all together, you get a very powerful style of mix.

For example, say you're mixing a love song. You could set the relative volumes to be "even" so that nothing jumps out and shocks you and the mood of the emotion is not disturbed. You could set the EQ to be natural so that nothing is too irritating in the midranges, everything is nice and bright, and there isn't too much bass to blow the mood. You could set the panning so that it is balanced (like a love relationship should be . . . yeah, right). You could use very few effects so that the mix is clean and clear (like your head is when you're in love). Or you could add some nice long delays and spacey reverb to simulate that floating feeling of being in love. You should refrain from creating any unnecessary movement with the faders, panpots, EQ, or effects so as to not spoil the mood. Using all of these tools together can create an intensely beautiful dynamic appropriate for the song. Such dynamics would be appropriate for any song that is beautiful, peaceful, calming, or just plain sweet.

On the other hand, say you're mixing some wild and crazy techno music in which the goal is to have fun and create some of the coolest, breathtaking effects possible in order to get people totally charged, excited, and up and dancing. You would set uneven volume relationships so that you would have extremely soft sounds followed by shockingly loud sounds. Let's wake some people up! Boost the low end to add some serious thump to the tush. Add some midrange on certain sounds in certain places to really cut through with a serious edge to get people's nervous systems going. Let's have some fun! Add some nice crispy highs to open the third eye. Now set the panning to be unbalanced, creating even more tension. ROCK AND ROLL!!! Then add a wide range of different effects, making the mix interesting at every single moment. Let's have some serious fun! Enough of the status quo. You could have things zooming left and right with panning, volumes going up and down, EQ changing throughout the mix, and effects and their parameters going up and down, as well as their parameters changing constantly. Using all of these tools together, you can create an intensely exciting dynamic that maximizes energy and fun and is totally appropriate for the song.

These are two extreme types of mixes that you can create with all of the tools and equipment in the studio, and all mixes in the world fall between these two extremes. Use these dynamics to spice up a song or to make it more beautiful. Again, do whatever is appropriate for the style of music and the thirteen aspects.

Changing the Style of Mix in the Middle of the Song

Even more intense is when you create a certain style of mix, then, using the four mixing tools, completely change all the parameters, creating an entirely different style.

The rock group Yes did it with "Owner of a Lonely Heart." They play a screaming electric guitar sound and, in a single moment, change to a 1950s style recording of a drum set, miked 20 feet away with a dull EQ. Then all of a sudden, they return to a screaming guitar/synthesizer type of sound that is extremely edgy. Sudden changes in multiple mix parameters can be extremely effective.

Sting also did it with the song "Englishman in New York." The song goes from a jazzy groove – very few effects, very clean, and small snare sound – to a huge drum sound with tons of reverb instantly. Then, in a flash, it is back to the simple, clear jazz mix.

Of course, you can only create such dramatic mix dynamics if they are appropriate for the song. This was obviously one of Frank Zappa's favorite techniques, and Mr. Bungle has taken this concept to the extreme; every 30 seconds, the song and the mix change completely.

To change the entire style of mix in a single moment can be shocking. It can blow people's minds. It can show people that their reality is just an illusion that can change at any moment. But best of all, it shows perspective. It shows people that they don't need to stay stuck in their current reality. They only need to put a different mix on the situation.

> All the mixes in the world
> are created with just these four tools:
> volume, EQ, panning, and effects.
> It's what you do with them that counts.

CHAPTER 7

Magic in Music, Songs, and Mixes

Now that we have discussed all the dynamics that can be created with the technical equipment in the studio, let's return to the basic concept:

> **The art of mixing is**
> **the way in which**
> **the dynamics you create with the equipment in the studio**
> **interface with**
> **the dynamics apparent in music and songs.**

Structuring the relationship between these two dynamics is the key job of the recording engineer. Herein lies the true art of mixing.

There is magic in the music and songs. There is magic in the way you turn the knobs on the equipment (the dynamics), and there is magic in the way that the two interface. The key is to be able to get really good at seeing the magic in each case. When it comes down to it, it's really about knowing what *magic* is in the first place, knowing what magic really feels like.

> The term "magic" is used to refer to whatever it is in music, songs, and the use of the equipment that turns you on. This could include beauty, intensity, massiveness, chaos, sincerity, love, or just *the flow* – whatever turns you on!

The trick is to develop a radar for finding magic. But it is a little more detailed than that. You must become totally immersed in this world of magic. You must become a master of the subtleties of magic. You must develop a highly refined hierarchy of magic. You need to categorize the magic you feel in your heart and entire being. Let's explore this a bit so you can see what I'm talking about.

First, what is the magic in music? Think about it. What are some examples of the magic you find in music? You should begin by developing your list. In fact, you probably already have a pretty long list, but you might not have really focused on it yet.

You can find magic in each of the thirteen aspects:

- **Concept** – one that touches you deeply.

- **Intention** – so strong you feel it in your Soul.

- **Melody** – you can't stop humming it and it makes you so happy or touches your heart deeply.

- **Rhythm** – just gets you moving.

- **Harmony** – so beautiful.

- **Lyrics** – helps you to see life from a different perspective. One that seems to fit exactly what is going on in your life, especially one that gives you a new perspective on a situation, or helps you get out of a rut or vicious circle of thought.

- **Density of arrangement** – arrangement that builds to a powerful peak and breaks down slowly to a profound state of peace and stillness. A perfect combination of simplicity (very few instruments) and complexity (many instruments) that doesn't bore you, but also isn't "too much."

- **Instrumentation** – just really cool instruments or sounds.

- **Song structure** – unique order of sections of the song.

- **Performance** – a performance that obviously took a huge amount of experience, practice, and talent to pull off. A singer that totally puts his/her heart and soul into the singing of the song.

- **Chills! Chills! Chills!**

- **Quality of recording and equipment** – well recorded with high quality equipment.

- **Mix** – cool effects.

- A **song** that just seems to fit your mood perfectly.

- A **combination of parts in a song** that seems to fit together in a way that creates something greater than the sum of the parts.

The question is, "What are examples of these aspects in music that light you up?" Think about the songs and music you like and try and pin down what the magic is that you really like.

Pay very close attention to the fine details of what turns you on in a particular piece of music or song. Some affect you one way, others affect you another way. Some are extremely deep, some are just cool. Some are fun, some touch your heart. Some are short in duration, some grow in you over a lifetime. Some touch you and your friends, some touch every soul in the world! There are various levels of feelings. It's like the difference between seeing the magic in a simple blade of grass, and then seeing the magic in a green field, or a range of green mountains. They all have their beauty, but the intensity of their beauty is different. Once you have different levels of magic clear in your head, you can then make judgments as what to highlight in a mix, or not.

The next step is to note exactly how the magic was created. Then, *name it*! Naming it helps you to pin down the idea and get it into your memory banks so you can use it later. Add it to your bag of tricks.

> Of course, magic is different for different people. You should not only learn what it is that turns you on, but also look for what turns others on. For example, whenever anyone says that they *really* like a song, I always pay close attention to figure out what it is that is lighting them up. Look for the magic that other people see.

Once you have found the magic, there are many ways to highlight it. You can simply turn it up in volume. You can add fattening to make it more present. Compression will also make it clearer. You might brighten it with EQ just a bit to make it stand out more. Just focusing your mind on the magic already creates a quantum effect where the listener's attention will also focus there (that's an interesting one, huh?).

> Really, when it gets down to it,
>
> what you are doing as an engineer
>
> is simply
>
> balancing out different levels of magic.
>
> It's like painting with magic.

The cool thing is that once you develop this radar for magic in music, it starts spilling over into the rest of your life. You start to develop an acute sensitivity for aesthetics in all that you see and do. Nature becomes more beautiful, any art becomes more intriguing to explore, and people become works of art to get into. Every great artist in the world develops radar for magic, an acute sensitivity for aesthetics.

> It's called being an "artiste."
>
> This is really what makes a great recording engineer.
>
> Anybody can learn what the equipment does,
>
> but only a few learn how to use it to create true magic.

First, you need to be totally familiar with all the magic the equipment can create. It is similar to being a great musician. Musicians live in a subtle world of feelings and colors. They are immersed in the magic of their craft. Once you become a master of the magic that can be created with the equipment, *and* are acutely aware of the magic going on in the music, the best way to use the equipment to enhance the magic becomes obvious.

Over time you will be able to just *feel* what works the best.

This is what makes a great recording engineer. The process doesn't happen overnight. But the journey is a pleasure.

> Don't miss the magic.
>
> And then, if there is no magic, you can sometimes use the recording equipment to create the magic.
>
> Or, if there is magic in the music,
> sometimes the magic created with effects might interface
> just perfectly to create a whole other level of magic –
> that might change people's lives forever.
> Hopefully.

Putting It All in Perspective

The recording engineer's job is to create musical dynamics with the equipment that reveals or enhances the magic people find in music. The trick is to use the dynamics created by the equipment to enhance, accentuate, highlight, support, create tension, or just let the music itself shine through (whichever is appropriate for the song and style of music). The way in which these dynamics interface with the music is the art of mixing. The art of the recording engineer is to seek out the relationships between the equipment and the music that are the best – whether they be magical, beautiful, amazing, world-changing, people-changing, or just cool. Don't stop until you get goosebumps and chills.

Using the visual framework, you can see how others utilize these possibilities to create great mixes. A whole new world of tricks and techniques are now at your disposal. You can now begin to explore all the different relationships between mixing dynamics and the dynamics that people perceive in music. You now have a framework to remember what you did, when you do something in a mix that totally works.

> Remember what you do – especially when what you do creates magic!
>
> After a couple of years of keeping track of the magic,
>
> you become a magician.
>
> Different people have different ideas of what great art is.
>
> The point is to develop your own values about what you think is great art.
>
> Then make it!

Now you know what is required of a recording engineer. Besides learning the technical side of the equipment and how to work well with a wide range of strange, unusual, and wonderful people, a recording engineer also deals with refining a diverse array of aspects – even the music itself. The mix is only one of many aspects that contributes to a great piece of recorded music.

You've learned many of the details of the dynamics that can be created with studio equipment. You now have a mixing framework, designed to include all of the musical possibilities, to help you get a good perspective on all that you can do in the studio.

Just as a great musician must, at some point, learn and incorporate aspects of theory and technique into his or her actual playing, so must the aspiring recording engineer incorporate theory into practice. This process varies for different people. Some people are able to take the theory and diligently incorporate it into reality in their mixes. For others you will find that a few ideas here and there will begin to sneak into your recording projects. You might also reread some sections of this book from time to time and gather a few more tidbits to incorporate. Regardless of your style of learning and rate of absorption, you now have an overall structure to see all that an engineer does, and you can focus on what it takes to become great at it.

Homework for the Rest of Your Life – Due (Do) Every Day

Whenever you have time to listen to a song with your full attention, try and pick out the magic in any one of the thirteen aspects of a recorded piece of music or in a combination of two or more of the thirteen aspects. Then ask yourself, "Why is it magical?" Often you may not even know. Second, ask yourself, "How did they obtain it? What did they do to be able to achieve this magic?" Then imagine what it would take for you to use it, and imagine how you might use it in a song.

CHAPTER 8

3D Sound Processors and Surround Sound Mixing

It is important to note that aurally, we are not very good at precise placement of sounds that are behind us. Therefore, the visual framework is very useful in displaying precise placement throughout a room.

This chapter might have been placed under "Panning" in Chapter 4. However, since this type of mixing is such a different animal we are devoting an entire chapter to it. The visual framework is especially useful when discussing the possibilities of 3D sound processors and surround sound mixing.

As discussed in Chapter 2, "Visual Representations of 'Imaging'," there is a limited space where a mix occurs between the speakers. When using a 3D sound processor or mixing in surround sound, this limited space is much larger.

A 3D sound processor uses phase cancellation and equalization cues to simulate the placement of sounds out in front of the speakers. Instead of the 3D space being here (as in normal stereo),

Visual 231.
Normal Stereo Field

it is expanded to here:

Visual 232.
3D Sound Processor
Stereo Field

With a 3D sound processor, the sound field will actually get larger as you step back from the speakers. The rear limit is always just about you! With a 3D sound processor, the illusion that is created only works when you are almost right between the speakers. If you are off to the side more than a few feet, you will not get the illusion. There are about five well-known companies that currently make 3D sound processors.

Surround sound normally uses five speakers and a subwoofer (which is why it is called 5.1 surround sound). Three speakers are placed up front and two in the back. With a surround sound setup, the world of imaging is expanded to here:

Visual 233.
Surround Sound:
Stereo Field

The obvious point with both 3D sound processors and surround sound mixing is that you have more space to work with. If you have a busy mix, you've got it made. In fact, I commonly utilize a 3D sound processor whenever I am working with a large number of instruments.

Since you have so much more space to work with, the density of the mix becomes a very important consideration. In a sparse mix with very few instruments, the amount of space between sounds can be overwhelming.

Visual 234.
Surround Sound:
Sparse Mix

In a denser mix with a large number of sounds, you simply spread things out more.

Whether you are using a 3D sound processor or mixing in surround, many of the general considerations are the same. Since somewhat of a new field, the traditions have not crystallized as much, although there are some trends that are becoming pretty serious.

A 3D sound processor is commonly used on a lead vocal to bring it more out front; Bonnie Raitt even used it on one album. It is also used on guitars to make them more in your face. I especially like using it when I have two or more guitars that are partially masking each other. You can bring one way out front and leave the other in the background to get almost total separation and still be able to hear both of them equally. I commonly bring out keyboards and strings as well.

One of my favorite effects is to put fattening on a sound, then send the fattening to the 3D sound processor and bring it out front. It makes it especially present. Then to create some serious spatiality, I send a stereo reverb to the 3D processor and pan it all around you.

One of the most effective uses of the unit is to take sounds and pan them around the room in real time. It is especially fun to take a repeating delay and swing it around your head.

When mixing in surround sound, all of these ideas translate perfectly; however, you have much more space to work with. With surround mixing, there are two types of mixes. The first attempts to preserve the normal traditions of stereo mixing and simply expands the stereo field. In this case, the rear half of the mix is normally reserved for (1) unusual special effects, (2) reverb (especially in a live recording), and (3) live audience miking.

The first type of surround mix is most commonly created in the front of the room with the rest of the room used for special effects or ambience. It normally extends out into the room from 1/4 of the entire space . . .

Visual 235.
Surround Sound:
Mix in the Front 1/4 to
1/2 of the Entire Space

Visual 236.
Surround Sound:
Mix in the Front 1/2

On rare occasions it will be extended to 3/4 of the room.

The second type of surround mix utilizes the entire space. This is quite common with dance music.

Visual 237.
Surround Sound:
Mix Throughout the
Room

In this case, there may not even be a front part of the mix. The entire space is considered to be a stage for sounds. Instead of the mix being mostly in front of you, you are completely immersed in it. This is especially the case when mixing for dance music, since the listeners might be anywhere, facing any direction in the room. A cappella music is often panned equally around the room.

Visual 238.
Surround Sound:
A Cappella Mix

Effects

Stereo placement of effects in a 3D world take on a whole new level of possibilities.

Following are some of the possibilities for fattening.

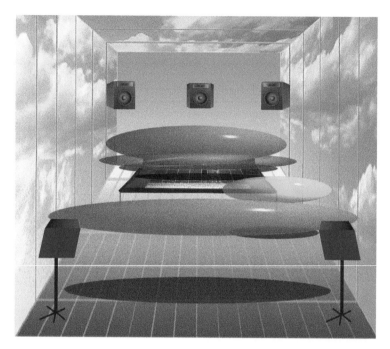

Visual 239.
Surround Sound:
Fattening Panned
Left to Right

Visual 240.
Surround Sound:
Fattening Panned
Front to Back

Visual 241.
Surround Sound:
Fattening Panned
Everywhere

Flanging, chorusing, and phasing may be panned in the same way as fattening. The possibilities for panning of reverb are also expansive.

Visual 242.
Surround Sound:
Reverb Panned
Everywhere

Visual 243.
Surround Sound:
Reverb Panned in the
Rear with a Predelay on
the Original Sound

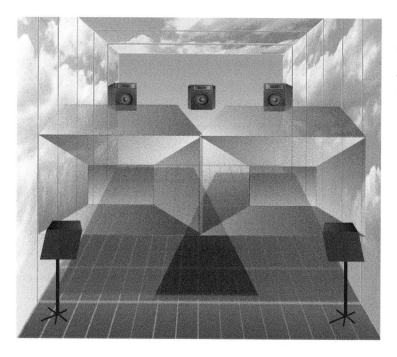

Visual 244.
Surround Sound:
Two Reverbs Panned
Left and Right

You can also get effects that are in 3D. In the reverb effects, each of the delays in the reverb is panned to a different spot in the room. When recording a live concert, it is especially effective to place mics throughout the room to pick up the natural reverb and pan the mics to the same positions in the surround mix. This gives the illusion of being at an actual concert.

Obviously, panning movement takes on a whole new level of excitement.

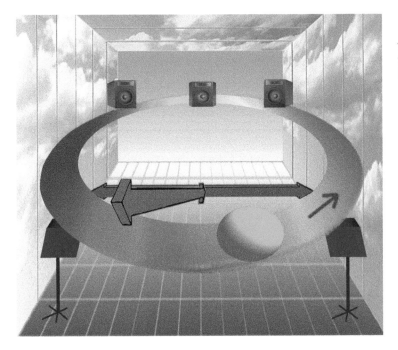

Visual 245.
Surround Sound:
Panning Movement

Let's take a look at how different instruments might be placed in the mix.

Vocals

Lead vocals are still most commonly placed front and center. There is an option to place the lead vocal in the center-front speaker, or to place it in both the front left and right speakers. Putting it in the front center speaker often makes it even more present; however, some listeners don't have a center speaker! As mentioned previously, if there is more than one vocal, they might be spread evenly throughout the room. Occasionally, a lead vocal is brought forward to make it feel more intimate.

Visual 246.
Surround Sound:
Lead Vocal Out
Front

Background Vocals

Background vocals are often placed left and right in the front speakers, but it is often a nice effect to bring them forward on the left and right.

Drums

Most engineers try to keep a live drum kit as cohesive as possible.

Visual 247.
Surround Sound:
Cohesive Drum Kit

Of course, this is not the case in any type of dance music.

Bass

As in a stereo mix, the bass is still mostly front and center, unless it is a high-end lead bass sound or performance. For dance music, it is most effective to place the bass in the center so all speakers are carrying the sound.

Visual 248.
Surround Sound:
Bass in the Center

Again, there are not as many traditions in surround sound mixing, so there is a lot of room for experimentation.

CHAPTER 9

Mixing Procedures

Now that you have a wider perspective on how it all works to create great mixes, let's go back to basics and explain how to build the mix from the beginning. This includes the mixing process, automation, and mastering.

Section A: The Mixing Process

Different engineers have their own procedures they follow when developing a mix. What follows is a process that will help you build the mix most efficiently.

Equalize Each Instrument Individually

You need to make each individual sound good; this means making the sound either natural or interesting. The primary considerations are mudd, low end, irritation, and brightness. If you have heard the entire song, you can also EQ the sound so it will sound good in the mix. Don't spend too much time working on a sound by itself – what the sound is like with the other instruments is what counts. Just get it in the ballpark, and again, if it already sounds good, don't touch it. Also, when in solo, always EQ it brighter than you think it should be. High frequencies are easily masked by the other sounds in the mix.

Bring Up the Mix

Bring Up Fader/Volume

The order and manner in which you bring up and balance out the levels of all the different instruments is important. It is helpful to establish and stick to a specific order. Don't bring up all your sounds at once. It is best to build a mix like a house – from the foundation up. Before you add the next sound, always make sure that your current mix sounds good. If you make sure it sounds good during every step of the building process, you will end up with a good overall mix. Also, if things start to sound weird, you can easily figure out at what point it went wrong.

Start by bringing up the kick drum about 2/3 of the way on the fader in order to leave plenty of room for a little extra boost later. Use your master volume (or amp) to set the desired volume in the room. Then use the volume of the kick drum as your anchor from which to build your mix. As you bring up the levels of each instrument, always listen first to what you already have and sneak in the next sound from underneath.

It is a good idea to build the foundation, or rhythm parts, of the mix first. Some engineers will bring up the vocals after bringing up the kick drum. Here's my preferred order to bring up your mix:

1. Drums: kick drum, snare drum, hi-hat, overheads, and toms
2. Bass guitar
3. Basic rhythm instruments: rhythm guitar, keyboard rhythms
4. Lead vocals
5. Lead instruments
6. Background and harmony vocals
7. Percussion

Pan to Taste

Pan each sound as you bring it up. Remember to conceptualize the mix before you bring things up. Think about how many sounds you have to spread between the speakers. As you bring up each section (e.g., drums and bass, main rhythm, leads, main vocals, and background vocals), consider the panning within the section and the panning of the section in relation to the rest of the sections. Some sections are mostly dictated by tradition (drums, vocals, and so on). However, the main rhythm section often offers a range of possibilities. Keep in mind the possible overall considerations: (1) a place where there is room for the sound, (2) natural panning as if you are there, and (3) asymmetrical versus symmetrical panning.

Add Effects

Many engineers will add effects as they bring up the instrument in the mix. However, you can't set the final level of the effects in solo because effects get masked by other sounds in the mix. You must always set the final level of effects while in the mix, with all sounds up.

Some engineers will bring up the entire mix before they add an effect to a sound. I recommend a combination – some effects now, some later. If a sound needs compression or gating, add it right when you bring it up. If I am planning to place a major effect on a sound – like flanging, fattening, long delays, or any 3D effect – I will bring it up as I bring up the sound itself. This is because a major effect can drastically alter what you do with the rest of the mix. If the effect uses up a large amount of your space, you will have to deal with it when you bring up the other sounds. Whenever you add effects to a sound, go ahead and make adjustments to the rest of the sounds in the mix if necessary. Again, you want the entire mix to sound good every step of the way.

Often, you need to make adjustments to the sound itself, as you put an effect on it. You might have to adjust the volume, panning, EQ, or effects to compensate for the addition of the effect.

Refine Volumes, Panning, Equalization, and Effects

Once you have your entire mix up, you should go through each instrument and refine the four settings of volume, panning, EQ, and effects. The question is this: do you start with one sound and check out the four settings? Or do you start with volume and go through all the sounds in the mix (then go through panning with each of the sounds, then EQ, and so forth)?

Normally, I choose a parameter (volume, pan, EQ, or effects) and go through each sound. However, say I'm going through all of the sounds and scanning to see if each of the volumes is precisely where I want it. If I find a sound that is too soft or too loud (based on everything we have already covered in the book – style of music, song details, and the people involved), I first ask myself, "Do the volume, panning, EQ, or effects have anything to do with the reason this sound is the wrong volume?" This often leads to a little detour where I will go and work on something other than the volumes.

Having completed checking the volume of each sound, you then checkout the relative EQ of all the sounds. This is the process in which you scan the highs, mids, and lows and make sure the wind isn't doing anything wrong. Again, if you find a sound that needs an EQ adjustment, first ask yourself if one of the other three components is contributing to the problem.

Scan the panning for each sound to see if there is anything that might need refining now that you have made some adjustments.

Scan the effects to see if you might need more or less effects, or if you might need a different effect.

Then, go back to your volumes.

The tricky part is when you adjust one setting, and that makes you want to adjust another setting, and that results in another change – it can seem to go on forever.

When you conceptualize a mix in the first place, and decide what you are trying to achieve, occasionally you have to make compromises for one value over another.

You continue with this round robin refinement process, until you're satisfied or almost out of time.

Section B: Automation

Once you have the four areas of the mix set, you'll often want to make changes for different sections of the song. There are four types of moves that can be done during a mix: volume, panning, EQ, and effects (volume, type, and parameters).

Automating Mutes

Most engineers will first record mutes. In a serious mix, most tracks are muted when there is nothing playing, especially if there is any type of noise at all on the track. This can be especially effective when a song starts out with only one or two sounds. By muting all of the other noisy tracks, you can make the mix sound totally clean and noise free. This is also the case at the end of the mix. You might mute (or fade down) sounds that might have noise in the background.

In a Digital Audio Workstation (DAW), the easiest way to muting is to simply select and delete the waveform on the track.

You can perform mutes by simply turning the mute button on or off while recording your automa (It's often nicer if you have a control surface rather than using a mouse). You can normally tell by looking at the waveform where the precise spot is. In fact, it is often preferable to begin by looking at the waveform and simply "cut out" the space between the waveforms by selecting the section with the mouse and hitting Delete on the computer keyboard. Since most DAWs are non-destructive, you can almost always drag out the edge of the waveform (un-trim) to be active again if you make a mistake.

Occasionally, it is necessary to fade a sound in or out using the fader instead of using the mutes to make the transition smoother.

Automating Volumes

As previously covered, volume moves include changing the volume of a sound at the beginning of a section or riding a fader throughout a section. Perhaps you need to turn up a guitar part for the lead section, or turn down the lead vocal for the chorus. These types of moves can easily be performed on the fader; however, it is often much easier to draw the moves using line automation while viewing the waveform. Then you can see precisely where the move should be performed, and you can easily change the volume of a whole section without having to wait for it to play through. This is also the case if you would like to raise the volume of an entire track throughout the song. Simply select the entire track (normally by double or triple clicking), and then you can grab the automation line with the mouse and raise the entire track volume at once.

When you are using a control surface, you can also ride the fader up and down while in record automation mode. As the volume gets too loud, immediately lower the fader, and as it gets too soft, raise it. Minor adjustments can then be made using the line automation section.

Overall fades of the master fader can also be automated. Unless the song fades in slowly, I will normally select the silence before the song begins (on all tracks) and delete it. Then when you *bounce to disk* (record the mix to two tracks), you will have a clean beginning. I will also normally automate the fade at the end of the song. Even if there is no fade, don't forget to fade the master fader after the last sound dies out.

An important point to remember when automating volumes is to always start from the beginning of the song. Besides being based on the volume of the other sounds in the mix, *the volume of any sound in a mix is always based on the volume of the sounds that came before it*. Therefore, start from the beginning. In particular be especially concerned with the volume of the vocal relative to the rest of the mix on the first word when it enters.

Often what happens when working with volumes is that once you get them in the ballpark, you start hearing finer and finer levels of volume differences. After getting all your volume levels adjusted, listen closer for finer refinements.

Automating Panning

Pan moves are less common in mixes. As mentioned previously, I am always on the lookout for an appropriate opportunity. Occasionally, you will simply make a panning adjustment in a section of a song. For example, you might pan a guitar track wider for a lead section, then return it to not so wide for the verse and chorus.

Panning is often mapped out using line automation while viewing the waveform for timing. If you are set to grid mode, you can also commonly use a pencil tool to create panning that is in time with tempo of the song.

Automating EQ

It is rare that you automate EQ. Occasionally, you will want to change the EQ of a section, or even use the EQ to fix a brief problem, such as rumble or a buzz. In electronica and dance music, sometimes the frequency knob is swept in time to the music. In most DAWs, these days, you can automate each and every parameter of the equalizer.

Sometimes it is simpler to copy the portion of the track over to another track and set the EQ differently on that track.

Automation Effects

The most common changes of effect parameters are to add a delay on the end of a line, line, line; and to adjust the amount of reverb or reverb time. Reverb is occasionally brought up in choruses, at the end of the song, or in special sections. Often I will adjust the amount of the fattening on a vocal when background vocals are added in a chorus. Occasionally, parameters of effects are automated. For example, you adjust the rate of the sweep on a flange to different speeds for different sections of the song.

As with EQ, most DAW plug-ins enable you to automate every single parameter. However, when automating the volume of an effect, you want to automate the auxiliary send that is sending the sound to the effect in the first place. When automating an effect on the end of a line, you want to automate the auxiliary send's mute. Set the level of the auxiliary send to the level you want, then automate the mute. You start with it muted, and then un-mute it on the last word of the line, line, line.

Section C: Mastering

(This section is a simplified version of mastering. There are whole books on mastering that are quite detailed.)

As previously discussed, the term *mastering* is used to mean many things.

Overall compression and limiting is often referred to as mastering, although the term often includes setting the order of the songs, the time between songs, EQ from song to song, as well as other aspects of noise reduction and even effects.

First, when you record the final mix to two tracks (bounce to disk), compress the overall mix. Set your ratio to 2:1 and set the threshold so you are getting approximately 3–6 dB of gain reduction. For more simple acoustic recordings, I set it to about 1–2 dB max. For solo instruments, I will often skip the mastering compression.

Listen carefully for any squashing of the sound. If you hear anything unusual, back off the threshold. Bring up the output of the compressor to compensate precisely for the amount of gain reduction. It is important to use a very high quality compressor. A multiband compressor that compresses different frequency ranges separately can also help the quality. Always keep the original because record companies often want something un-mastered so they can use a top mastering facility and engineer.

After you get all of your songs for the CD pre-mastered, create a whole new song file and bring them all back into your computer. If the songs have an even EQ overall, place them on the same two tracks end to end. If you decide that one song needs to be EQ'd differently than another song, it is much easier to cut and paste it to two new tracks instead of trying to automate the EQ. Also, if you want to do a crossfade, you can put the second song on its own two tracks.

Listen to each song to make sure there are no digital glitches. Then, set the order of the songs. The order can be based on a variety of considerations. First, try different orders and simply set it completely on whatever feels right. This is the most highly recommended. If you are doing a demo for a record company, be sure and put the songs in the order from best to worst. Otherwise, I will normally use one of the best songs to start the album, and try to get the very best song to be within the top three songs on the album. It's also nice to save a really good song for the end, in order to leave the listener playing it over and over in his brain. Again, do whatever feels the best.

Set the time between songs based on your feelings. Only use preset times between songs when in a serious time pinch. Listen to see if the time feels good. Listen to the last 10 or 15 seconds as the song fades out and note where it feels like the next song should start. Often it feels better to have a longer time between songs after a song with a slower tempo. Sometimes you might want no time between the songs, or you might want a crossfade. Check it out and set it however, you like it because if you don't, the wind will. And you know the wind . . .!

Next, if the files aren't recorded hot enough, normalize the files to get them up to snuff. Then, set the volumes from song to song. Don't use the meters. Use your ears. Listen to a few seconds in the middle of the song, then click on the next song and listen to a few seconds in the middle of it. Go back and forth until they sound somewhat even. Then double-check it by listening to the first song's fade and compare it to how loud the next song comes in. Again, set it so it feels right.

Next, set the overall equalization. There are two considerations here. First, and most importantly (and as explained previously), you *must* get the overall EQ correct. Listen on as many pairs of speakers as possible. As previously mentioned, I used to take my recordings to a local stereo store to listen to them on a wide range of speakers. Listen to it on headphones. Listen to it in your car. Listen to it on a small boom box. Also, compare it to a CD of a similar style of music.

If all of the tracks are on the same two tracks, it is then easy to pull up an equalizer on those tracks and set them. If you have songs on different tracks, then buss all your tracks to an auxiliary channel and place the overall EQ on that channel. In most software programs, this means simply adding a master fader channel and adding the EQ to it.

The second consideration with EQ is the overall EQ from song to song. As mentioned, it is easiest to simply place each song that needs its own EQ on its own two tracks. This way you can use the individual tracks to EQ each song on its own and the overall EQ for all of the songs (if necessary). This is especially important when doing a compilation album on which the EQs of each song vary drastically.

Some other things that you might consider include noise reduction, rearranging or shortening sections of songs, overall effects (perhaps only in certain sections of the song), checking for phase coherence, and 3D processing of the whole stereo mix.

After having completed all of the above, add a mastering limiter to your master fader channel or whatever channel you have bussed everything to. Set your ceiling so that it is between 0.1 and 0.5 dB. I normally set it to negative 0.3 dB. Then set your threshold so you are getting a maximum of 1–2 dB of gain reduction. Listen very closely for any squashed sound – especially on the very loudest sections of the songs. On most mastering limiters, when you bring down the threshold, it automatically raises the output, so there is no need to set the output.

You should now be able to play back the song file from beginning to end, and it should sound like a finished CD. Now you need to bounce each track to disk again *with the times between songs already included*. Select the first song, including the space between the first two songs, up to just before the second song, and bounce it to disk. If you have a crossfade, select the first song up to just before the beginning of the second song – even though this won't include the complete ending of the first song. On the CD, it will play fine. This way you will be able to select the beginning of the second song on your CD player when you are playing the CD.

Once you have bounced all the songs to disc, simply throw them into your CD pressing software in the correct order. Make sure you set the time between the songs to zero because you have already bounced them to disk with the space between the songs! This will also make any crossfades seamless. Listen to the CD to make sure there were no bugs during the pressing. I will listen to at least the first 15 seconds or so if I am short on time.

Using an Equalizer

Outline of the Step-by-Step Process

(To see the complete detailed process go to "The Step-by-Step Process for EQ'ing a Sound" in Chapter 4, page 79)

1. Pull up the equalizer and reset the volume controls to "0" if necessary.

2. Listen for

    a. Mudd (100–400 Hz)

    b. Low bass (40–60 Hz)

    c. Irritating frequencies in the midrange (800–5,000 Hz)

    d. Brightness (3,000–8,000 Hz)

3. Set the bandwidth.

    Thin for mudd, low bass, and irritation.

    Medium for brightness.

4. Find the frequency to be boosted or cut.

    Boost volume all the way (or almost all the way).

5. Sweep the frequency knob to find the frequency to be boost or cut.

6. Return the volume knob to "0" to regain consciousness.

7. Boost or cut the volume to taste.

8. Check it to see if you like what you did; turn the EQ switch on and off to compare.

Frequency	40–100	100–400	400–1000	1000–5000	5000–8000	8000–20,000
Sounds						
Bass	Bottom	Roundness, Muddiness	Body on Small Speakers	Highs, Presence		Hiss
Kick	Bottom	Roundness, Muddiness		Thud	Click	Hiss
Snare	X	Fullness, Muddiness			Brightness, Clarity	
Toms	X	Fullness, Muddiness		Presence	Brightness, Clarity	
Floor Toms	Bottom	Fullness, Muddiness		Presence	Brightness, Clarity	
Hi-Hat, Cymbals	Bleed	Fullness, Muddiness, Bleed		Irritation	Clarity, Crispness	Shimmer, Sizzle
Voice	Rumble	Fullness, Muddiness	Honkiness	Presence, Irritation, Telephone	Clarity, Crispness, Sibilance (6K)	Sparkle
Piano	Bottom	Fullness, Muddiness	Honkiness	Presence	Brightness, Clarity	Harmonics
Harp	Bottom	Muddiness, Pedal Noise,			Brightness, Clarity	
Electric Guitar	X	Fullness, Muddiness, Crunch		Twanginess, Cut/ Shred, Irritation	Thinness	Hiss
Acoustic Guitar	X	Fullness, Muddiness			Brightness, Clarity	Sparkle
Organ	Bottom	Fullness, Muddiness		Presence	Brightness	
Strings	X	Fullness, Muddiness		Irritation, Scratchiness	Clarity, Crispness	Shimmer
Horns	X	Fullness, Muddiness	Roundness		Clarity, Crispness	
Conga	Boominess	Fullness		Presence	Clarity, Crispness	
Harmonica	X	Fullness		Presence		

Chart 4. Equalization Chart

Mastering

Outline of the Step-by-Step Process

(To see the complete detailed process go to "Mastering" in Chapter 9, page 261)

Overall compression and limiting is often referred to as mastering, although the term often includes setting the order of the songs, the time between songs, the EQ from song to song, and other aspects of noise reduction and even effects.

1. Compress the overall mix at a 2:1 ratio and set the threshold so you are getting approximately 3–6 dB of gain reduction. Listen carefully for any squashing of the sound.

2. Import all pre-mastered songs into a new song file. Place them end to end. Decide whether you need them on different tracks for different EQ and crossfades. Otherwise, put them all on two stereo tracks.

3. After you get all of your songs for the CD pre-mastered, create a new song file and bring them all back into your computer. If the songs have an overall even EQ, place them on the same two tracks end to end. If you decide that a song needs to be EQ'd differently than another song, it is much easier to cut and paste it to two new tracks instead of trying to automate the EQ. Also, if you want to do a crossfade, you can put the second song on its own two tracks.

4. Listen to each song to make sure there are no digital glitches.

5. Set the order of the songs.

6. Set the time between songs.

7. Normalize, if necessary, to maximize the recorded volume.

8. Set the volumes from song to song. Don't use the meters – listen.

9. Set overall equalization for each song, and overall if necessary.

10. Other considerations: noise reduction, rearranging, or shortening sections of songs, overall effects (perhaps only in certain sections of the song), checking for phase coherence, and 3D processing of the whole stereo mix.

11. Add a mastering limiter. Set ceiling to negative 3 dB. Set threshold so you are getting a maximum of 3 dB of gain reduction. Listen very closely for any squashed sound, especially on the very loudest sections of the songs.

12. Bounce each song to disk with the blank spaces between songs.

13. Throw all of the mastered songs into a CD pressing program and make sure it is set so there is no time between songs. Press your CD.

14. Check your CD for errors.

Homework for the Rest of Your Life –
Due (Do) Every Day

An Overview of All Homework

Critique the Thirteen Aspects

Quality is defined in different ways by different people, so it can take a while to learn all the ways in which songs can be refined. When it comes to values, the only one that's really bad is "no values at all." To develop your own values, start focusing on each of the thirteen aspects whenever listening to music. As you check out the details of each of these components in the songs you listen to, you will develop a range of values. Even if you know nothing about "music theory," you can still learn how to make careful suggestions about each of these components. Then you will have more to offer during the recording session.

Whenever you have time to listen closely to a song, critique each one of the thirteen aspects. Try and define what the engineer or producer did for each aspect. Then ask yourself, "Do I like this or not?" At first, the answer to this question might be, "I don't know." However, if you simply start paying close attention to each component, you will naturally develop your own perspective on what you like and what is "good." Meanwhile, many of you have already developed some pretty detailed ideas as to what quality means for each aspect. In fact, it seems, some people are born with it. From years of teaching, I have found that many people already have very specific and highly developed values, but they often have never articulated them. So the trick is to define them and put the values into words. Not only does this help you to remember the values, you will be more confident to share them in recording sessions, whenever appropriate.

Another good exercise is the following: whenever you hear a song that you really like, ask yourself why you like it. Which one of the thirteen aspects is it that makes you like the song? It might be a combination. This way you begin to pin down your own values for each of the thirteen aspects. Inevitably, your values will not only start to shift, they will also become deeper and more refined.

It is good to get into the habit of critiquing each of the thirteen aspects on every song that you have the time to listen to in detail. The truth is, this is what professional engineers and producers do all the time. At first, it is tedious. Later it becomes second nature. Ultimately, it helps enrich your listening pleasure because you are able to get more depth out of the music you listen to. But most importantly, it expands and deepens your range of values so you have something to offer in a recording session.

Listen to Other Styles of Music

If you have a favorite style of music that you listen to all the time, start listening to other types on the radio. Listen for the differences in the mixes for each style of music.

Which of the Thirteen Aspects Is the Mix Based on the Most?

Every time you listen to a song (and have time to listen), check it out and see which one of the thirteen aspects is playing the most important part in the creation of the mix. Over time, you will start detecting patterns for different styles of music and different songs.

Amount of Bass on Kick Versus Bass on Bass

Have you ever paid attention to the difference in the amount of bass EQ on the bass guitar compared to the bass EQ on the kick drum? Which should have more? The truth is that it depends on the style of music and the song itself (and sometimes on the opinions of the bass player and drummer). For example, in reggae and blues, the bass often has more low-end. The kick (especially the 808 boom) in rap often has the most low-end. Start checking it out on songs from now on, and see what others are doing. And don't forget to ask yourself if you like what they did. Very soon, you will develop your own values as to how much bass you like on the kick versus the bass for different styles of music and songs.

Critiquing Compression Levels

Pay attention to the overall amount of compression that seems to be going on in each song you hear and develop your own values for how much compression you like.

Critiquing All Four Tools on Each Instrument

Every time you have the time to listen closely, check out every single sound in the mix, and see where it has been placed as far as volume, panning, EQ, and effects. Then for each sound ask yourself the first question:

Why might they have placed it there?

Take volume for example – determine the relative level of every sound in the song on a scale of 1–6. Listen to the volume level of each instrument and ask yourself, "Why might the engineer have put it at the level it's at?"

Of course, this is a game. You never really know. However, as you try to figure out the real reasons that engineers use to place sounds where they are, you then develop a good perspective on all the reasons.

Some of the most common reasons are:

1. That's the way it is normally done in this style of music.

2. Something about one of the eleven aspects caused the engineer to place at that level.

3. Somebody in the room asked for it to be at that level.

4. They were either nuts or on drugs.

Again, this helps you to learn all of the considerations that affect what you do in a mix with each of the four tools. More importantly, it gives you the *real reasons* that you can give to someone who is telling you to make the mix suck. This exercise helps you to develop your repertoire of reasons to give people to explain why things should be the way they should be.

Then ask yourself the second question:

Do I like it there?
(Volume, panning, EQ, and effects for each sound)

At first, you may not have any preferences. The answer to this question might very well be, "I don't know. A little louder, a little softer; pan left or right; EQ more bass or treble; more or less effects . . . I don't really know." But if you simply pay attention to the levels and mix settings of each sound, after awhile, you end up with a very detailed perspective on what is being done in mixes. You will then develop your own values, and you will know exactly where you like the levels and settings of different instrument sounds for various styles of music and songs. Then when you go into the studio, you'll no longer be unsure about where to set your volumes, panning EQ, and effect. You not only know what everyone else is doing, you know what you like in relation to the rest of the world. You will have developed your own style. Then you are a professional engineer.

The best part of it all is that other people will see that you know. Ultimately, you will gain their respect. Then, when they tell you to make a mix suck, all you have to do is slowly turn around and look them in the eye. Often you don't have to say anything – they will know.

By the way, all professional engineers do this exercise all the time.

Where Is the Magic?

Ask yourself where the magic is in every song you listen to. Define it, and try and figure out how they did it.

Whenever you have time to listen to a song with your full attention, try and pick out the magic in any one of the thirteen aspects of a recorded piece of music or in a combination of two or more of the thirteen aspects. Then ask yourself, "Why is it magical?" Often you may not even know. Second, ask yourself, "How did they obtain it?, What did they do to be able to achieve this magic?" Then imagine what it would take for you to use it, and imagine how you might use it in a song.

The Virtual Mixer™

Patent #5,812,688

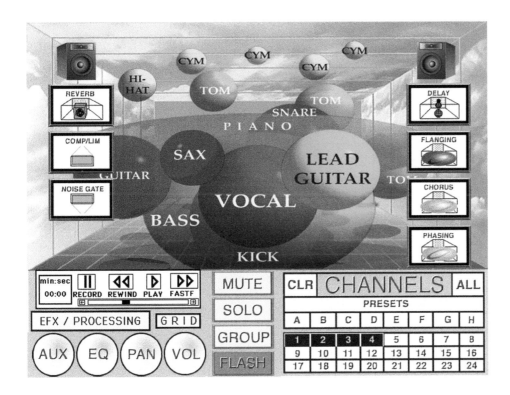

A 3D Visual Interface for Displaying Mixes and Controlling All Mixing Functions

Besides using this visual framework as an educational tool, it is now being developed as an interface for mixing consoles and effects. The interface uses three-dimensional representations of visuals of sounds between the speakers instead of two-dimensional pictures of a mixing board. It is intuitive and gives engineers (and students of engineering) more information about the mix, such as masking. The interface also works with 3D glasses for an enhanced visual presentation, and it is particularly useful for surround sound mixing.

Background

Over 20 years ago, we came up with the specific framework for mapping audio parameters into the visual world. The interface was originally developed as a tool to explain various styles of mixes, but then we realized it could be used to control a console. At that time, computers were not fast enough, and there were no MIDI-controllable consoles. The system is now being developed as an interface for digital consoles, and digital audio workstations.

What It Does

The Virtual Mixer is a graphical user interface that displays the amplitude and frequency range of sounds on a multitrack. It then allows you to use these visuals to control various types of mixing consoles and effect units.
The Interface does two things:

1. It is used as a new type of "metering" system to display the audio information of sounds in a mix. Amplitude (same as the meters on a multitrack) is shown as a function of brightness, which makes the images flash to the music.

2. Movement of the visuals sends out MIDI information (or other automation protocols) to the console and effects to control all parameters.

Basics

Fader level (volume) is mapped out as a function of front to back so that louder sounds appear closer (and, therefore, a little larger), and softer sounds appear more distant. The grid on the floor is calibrated to the faders on the mixing console. Panning is naturally mapped out as a left to right placement. The average pitch of the sound over the entire song is mapped out as a function of up and down. Higher pitch sounds appear higher between the speakers; lower pitch sounds appear lower, just as they do when you focus on imaging. Also, high-frequency sounds are smaller because they take up less space in a mix. Low-frequency sounds are larger because they mask other sounds more in a mix. You can also bring up windows that show equalization and auxiliary send levels visually.

The next step is to put on 3D glasses to see the images floating in 3D right where the sounds seem to be between the speakers. In such a virtual environment, the mixer can use a dataglove to move the sounds around in the mix. If you want a sound louder, pull it toward you. If you want it in the left speaker, put it there. You can even toss sounds back and forth. With surround sound, the images are floating all around you – throughout the room. You can even bounce sounds off the walls. *Then with a Sound Chair, you can actually place sounds inside your body!!!*

Formats

The interface is being developed for a wide range of mixing equipment: digital consoles, sequencers, and digital audio workstations. We will also be putting out versions that will control various multi-effects units and will display specific synthesizer sounds as texturized spheres corresponding to the waveform.

Pricing

The basic interface for a digital console or workstation will be around $300–$500. A touch-sensitive screen add-on will be around $500. A full 3D version (with 3D glasses on a computer monitor) would be under $1,000. Full virtual reality with head tracker display helmet will be around $1,500. Full big-screen projection will be around $10,000. Actual prices will drop as demand for the interface increases and as 3D technology continues to come down in cost.

More Engrossing

When you see the mix visually, you are incorporating more of your senses into the musical process. The more senses you use (especially in a fleeting artistic endeavor such as mixing sound), the more your entire being is engulfed in the experience. And the more you are engrossed in all of the details, the easier it is to come up with a great mix.

More Intuitive

Pictures of sounds are one logical step closer to the music you are mixing than faders and knobs on a console. Manipulating visuals of the sounds themselves is much more intuitive than pictures of a console. Pictures of sounds are more like music than knobs.

More Conducive to the Creative Process

Studio equipment is notorious for getting in the way of the creative process. Faders and knobs on a console distract recording engineers from the music they are trying to mix. Images that flash to the music help the user to focus more clearly on the invisible sounds they are mixing.

More Information About the Mix

The visuals provide additional information helpful for the mix. The primary goal of the visual system has been to show "audio masking" visually. Engineers can use this information to discover hidden problems, and best of all, to be able to explain the problem to the band and producer. For example, a bad arrangement becomes clearly evident.

More Perspective

The visual framework provides the engineer with a perspective on all of the possibilities available to him. The framework shows each parameter within each piece of equipment in the studio and shows how each parameter contributes to an overall mix. With this visual perspective of all that is possible in a mix, a humongous number of possibilities are displayed in a way that puts an array of creative ideas at an engineer's fingertips.

More Relationships

The interface shows more than just all of the settings of each piece of equipment in the studio. The interface shows the *relationships* of all the settings and how they work together to create a mix. After all, it is the relationships of all the settings that really count.

Better Communication

The interface enhances and simplifies the communication process between recording engineer and client (band or producer). The interface is so intuitive that even inexperienced clients will be able to follow the development of the mix and communicate effectively with the engineer.

More Fun

And of course, we mustn't forget – flashing 3D visuals are a blast to watch and work with.

"THE NATURE OF THE MEDIUM AFFECTS THE ART YOU CREATE."

The Virtual Mixing Company
351 9th St. #202, San Francisco, CA 94103
David@GlobeRecording.com
415 777–2486
www.VirtualMixer.com

E-mail (or mail) us if you might be interested, and we'll let you know when it is available.

Index

Page numbers in *italics* show visuals, **bold** a chart